Progress in Mathematics
Volume 79

Ralph McKenzie   Matthew Valeriote

Structure of Decidable Locally Finite Varieties

1989

Birkhäuser
Boston · Basel · Berlin

Ralph McKenzie
Department of Mathematics
University of California
Berkeley, California
U.S.A.

Matthew Valeriote
Department of Mathematics and Statistics
McMaster University
Hamilton, Ontario
Canada

Library of Congress Cataloging-in-Publication Data
McKenzie, Ralph,
The structure of decidable locally finite varieties / Ralph
McKenzie, Matthew Valeriote.
p. cm. -- (Progress in Mathematics ; v. 78)
Includes bibliographical references.
ISBN 0-8176-3439-8
1. Varieties (Universal algebra) I. Valeriote, Matthew.
II. Title. III. Title: Locally finite varieties. IV. Series:
Progress in mathematics (Boston, Mass.) ; vol. 78.
QA251.M44 1989
512--dc20

Dr. McKenzie's research is supported by National Science Foundation grant number DMS-8600300.
Dr. Valeriote's research is supported by the Natural Sciences and Engineering Research Council of Canada.

Printed on acid-free paper.

ISBN 0-8176-3439-8
ISBN 3-7643-3439-8

Text prepared by the authors in camera-ready form.
Printed and bound by Edwards Brothers, Ann Arbor, Michigan.
Printed in the United States of America.

9 8 7 6 5 4 3 2 1

This book is dedicated to our wives --
Kathie and Theresa

Contents

Introduction

A mathematically precise definition of the intuitive notion of "algorithm" was implicit in Kurt Gödel's [1931] paper on formally undecidable propositions of arithmetic. During the 1930s, in the work of such mathematicians as Alonzo Church, Stephen Kleene, Barkley Rosser and Alfred Tarski, Gödel's idea evolved into the concept of a recursive function. Church proposed the thesis, generally accepted today, that an effective algorithm is the same thing as a procedure whose output is a recursive function of the input (suitably coded as an integer). With these concepts, it became possible to prove that many familiar theories are undecidable (or non-recursive)—i.e., that there does not exist an effective algorithm (recursive function) which would allow one to determine which sentences belong to the theory. It was clear from the beginning that any theory with a rich enough mathematical content must be undecidable. On the other hand, some theories with a substantial content are decidable. Examples of such decidable theories are the theory of Boolean algebras (Tarski [1949]), the theory of Abelian groups (Szmielew [1955]), and the theories of elementary arithmetic and geometry (Tarski [1951], but Tarski discovered these results around 1930). The determination of precise lines of division between the classes of decidable and undecidable theories became an important goal of research in this area.

By an *algebra* we mean simply any structure $\langle A, f_i(i \in I)\rangle$ consisting of a nonvoid set A and a system of finitary operations f_i over A. A *variety*, or *equational class*, is a class of similar algebras defined by some set of equations. A variety is called *locally finite* if every one of its finitely generated algebras is finite. A variety is called *decidable* if and only if its first order theory is a recursive set of sentences. In this book we address the questions: Which varieties are decidable? If a variety is decidable, what can one conclude about the structure of its algebras?

W. Szmielew proved in [1955] that every axiomatically defined class of Abelian groups is decidable. Yu. L. Ershov [1972] proved that every variety of groups containing a finite non-Abelian group is undecidable. A. P. Zamyatin, in a series of papers published between 1973 and 1978, showed that every non-Abelian variety of groups is undecidable, and went on to characterize all the decidable varieties of rings and semigroups. (See the bibliography.) Inspired by Zamyatin's success, S. Burris and R. McKenzie [1981] considered varieties that were completely unrestricted in nature, except that they had to be locally finite and congruence-modular. Using Zamyatin's methods and some new techniques, they reduced the problem of determining the decidable varieties, within this domain, to two much more restricted problems: Which varieties of modules over finite rings are decidable, and which discriminator varieties are decidable?

In the present work, we remove the hypothesis of congruence-modularity from these results. We prove that a decidable locally finite variety is the product of a decidable, congruence-modular variety, and a decidable, strongly Abelian variety; and we establish a simple criterion for the decidability of a locally finite, strongly Abelian variety. We find that there are just three kinds of indecomposable, decidable, locally finite varieties, characterized by extremely different structural features of their algebras.

Varieties of the first kind are *strongly Abelian*. A decidable variety of this kind is definitionally equivalent with a class of k-sorted multi-unary algebras for some integer k. Valeriote [1986] found a simple necessary and sufficient condition for the decidability of a locally finite, strongly Abelian variety, which is reproduced here in Chapters 11 and 12.

Varieties of the second kind are *affine*. Each variety of this kind is equivalent, in a very strong fashion, to the variety of modules over some finite ring. The problem of determining which locally finite affine varieties are decidable is equivalent to that of determining for which finite rings the variety of modules is decidable—an unsolved problem. Some partial results on this problem are mentioned in Chapter 14.

Varieties of the third kind are *discriminator varieties*. The generic discriminator variety is the (decidable) class of Boolean algebras. The problem of characterizing the decidable locally finite discriminator varieties is open. However, it is known that each discriminator variety which, like the variety of Boolean algebras, is generated by a finite algebra with finitely many basic operations, is decidable.

This whole book consists of a proof of one result, Theorem 13.10. The theorem asserts that every decidable locally finite variety $\mathcal{V}$ is the product (or join) of independent decidable subvarieties, $\mathcal{V}_1$, $\mathcal{V}_2$ and $\mathcal{V}_3$, of the three kinds described above. The independence of these subvarieties means that every algebra in $\mathcal{V}$ is uniquely expressible as the direct product of three algebras, one from each of the subvarieties. The proof shows that if a locally finite variety does not decompose in this way, as the join of independent varieties of the three special kinds, then at least one of twenty different interpretations will interpret the class of all graphs into the variety. Fifteen of these interpretations are developed in this book, while the other five can be found in texts to which we refer.

The research leading up to this result was motivated only in part by a desire to know which varieties are decidable. As algebraists, we have a compulsion to reveal and describe the structural features of algebras. We hoped that the study of decidability might reveal some rather precise division of the family of all varieties into a small number of subclasses composed, on the one end, of varieties in which structure is manageable in all its aspects, and can be described, and on the other, of varieties that sustain structures of arbitrary complexity. In fact, we view our chief result, Theorem 13.10, as an important step in this direction; and it seems to indicate that the connection between decidability and manageable structure is a very close one.

The book begins with a preliminary chapter in which we introduce the concepts with which we shall be working, and the tools that we shall use. Next, the long argument that proves Theorem 13.10 is presented in Chapters 1–13. The structure of the argument and the plan of these chapters are outlined in the first paragraphs of Chapter 1. In this work, the modified Boolean powers introduced in Burris, McKenzie [1981] continue to play a substantial part; while a new ingredient, the tame congruence theory of D. Hobby, R. McKenzie [1988], has an indispensable role.

As a corollary of Theorem 13.10, we have an algorithm which leads from any finite algebra with finitely many basic operations to a finite ring with unit, so that the variety generated by the algebra is decidable if and only if the variety of unitary left modules over the ring is decidable. A second corollary is that any finite algebra contained in a decidable variety generates itself a variety that is definitionally equivalent with a finitely axiomatizable, decidable variety. These corollaries, and four open problems, are discussed in Chapter 14.

This book is divided into three parts. All of the results in Part II, as well as Lemma 13.12, are drawn from Valeriote's [1986] doctoral dissertation. The authors are grateful to Stanley Burris and Ross Willard for a very careful and critical reading of the manuscript. The improvement owing to their suggestions is very visible to us, especially in Chapters 2 and 6. We would also like to thank Bradd Hart for the helpful suggestions he provided on the presentation of Chapter 11.

Chapter 0

Preliminaries

For more detail on the topics briefly introduced here, the reader can be referred to: S. Burris, H. P Sankappanavar [1981] for universal algebra, lattice theory, first order logic and decidability; S. Burris, R. McKenzie [1981] for several theorems quoted here; and D. Hobby, R. McKenzie [1988] for a detailed development of tame congruence theory, including many results that will be applied here. The introductory chapters in Burris and McKenzie contain an introduction to decidability, interpretations, discriminator varieties and affine varieties that complements what we write here. The proofs in Burris and McKenzie relied on several results about commutators in congruence-modular varieties that had not appeared in print at that time. The reader was referred for these results to a preprint by R. Freese and R. McKenzie titled *The commutator, an overview.* That paper has now been published (much expanded) as Freese, McKenzie [1987].

0.1 Languages, structures, algebras and graphs

A **first order language** L consists of a set Φ of operation symbols, a set Σ of relation symbols (disjoint from Φ), and a function ρ that assigns to each member τ of $\Phi \cup \Sigma$ a non-negative integer $\rho(\tau)$ (called the *rank* or *arity* of τ). Formally, we have $\mathsf{L} = (\Phi, \Sigma, \rho)$. L is an *algebraic* language if Σ is empty; a *relational* language if Φ is empty. A **model** of a first order language L, or **L-structure** is a system

$$\mathbf{A} = \langle A, f^{\mathbf{A}}(f \in \Phi),\ s^{\mathbf{A}}(s \in \Sigma)\rangle$$

consisting of a non-void set A, a $\rho(f)$-ary operation $f^{\mathbf{A}}$ on A associated to each $f \in \Phi$, and a $\rho(s)$-ary relation $s^{\mathbf{A}}$ over A associated to each $s \in \Sigma$. The class of all models of L is denoted Mod (L). Members of this class are called **algebras** if L is algebraic, or **relational structures** if L is relational. The **type** of a structure $\mathbf{A}$ is its language, which we will denote by $\mathsf{L}(\mathbf{A})$. Thus a structure $\mathbf{A}$ of *finite type* has just a finite number of operations and relations, while $\mathbf{A}$ is *finite* if its *universe*, A, is a finite set.

In the considerations of tame congruence theory, it is natural to admit another kind of algebraic system (see §0.6 below), namely a pair $\langle A, F\rangle$ in which A is a nonvoid set and F is a set of finitary operations on A. Algebraic systems of this second kind will be called **non-indexed algebras** (or sometimes just *algebras*), and where it it necessary to emphasize the distinction, algebras that are models for some language will be called **indexed algebras**.

The set $\{0, 1, \ldots, n, \ldots\}$ of natural numbers is denoted ω; and A^n (where n is a natural number) denotes the set of all sequences, or n-tuples,

$$\langle a_0, a_1, \ldots, a_{n-1} \rangle$$

with $\{a_0, \ldots, a_{n-1}\} \subseteq A$.

Given a set A and natural numbers $n > 0$ and $i < n$, we can define the projection function $p_i^n : A^n \to A$ by

$$p_i^n(x_0, \ldots, x_{n-1}) = x_i \text{ for all } x_0, \ldots, x_{n-1} \in A.$$

A **clone** on a set A is a set of finitary operations on A that contains the projections p_i^n for all $0 \leq i < n < \omega$, and is closed under composition of operations. By a **polynomial clone** on A, we mean any clone on A that also contains all of the constant 0-ary operations on A. (A general clone is not required by our definition to contain any 0-ary operations.)

Associated with any algebra $\mathbf{A} = \langle A, f^{\mathbf{A}} \, (f \in \Phi)\rangle$ are two important clones. The **clone of term operations** of $\mathbf{A}$, denoted Clo $\mathbf{A}$, is the clone on A generated by the set $\{f^{\mathbf{A}} : f \in \Phi\}$ of basic operations of $\mathbf{A}$. An n-ary operation f on A belongs to Clo $\mathbf{A}$ iff there exists a term t in the language L such that $f = t^{\mathbf{A}}$—i.e., f is the operation induced by the term t. The **clone of polynomial operations** of $\mathbf{A}$, denoted Pol $\mathbf{A}$, is the clone on A generated by the basic operations of $\mathbf{A}$ along with all of the constant 0-ary operations on A. We write $\mathrm{Clo}_n\mathbf{A}$ for the set of all n-ary members of Clo $\mathbf{A}$ (n-ary term operations of $\mathbf{A}$); and, similarly, we write $\mathrm{Pol}_n\mathbf{A}$ for the set of all n-ary members of Pol $\mathbf{A}$ (n-ary polynomial operations of $\mathbf{A}$). It follows that if $p \in \mathrm{Pol}_n\mathbf{A}$, then for some $m \in \omega$ and $a_0, \ldots, a_{m-1} \in A$ and for some $f \in \mathrm{Clo}_{m+n}\mathbf{A}$, we have $p(\bar{x}) = f(\bar{x}, \bar{a})$ for all $\bar{x} \in A^n$ (where $\bar{a} = \langle a_0, \ldots, a_{m-1}\rangle$).

A **subuniverse** of an algebra $\mathbf{A}$ is a subset of $\mathbf{A}$ that is closed under each of the basic operations (and, hence, under each term operation) of $\mathbf{A}$. A **congruence** on an algebra $\mathbf{A}$ is an equivalence relation on the universe A of $\mathbf{A}$ that is also a subuniverse of $\mathbf{A} \times \mathbf{A}$. The subuniverse of $\mathbf{A}$ generated by a set $X \subseteq A$ will be denoted $\mathrm{Sg}^{\mathbf{A}}(X)$; it is identical with the set

$$\{t(x_1, \ldots, x_n) : n \geq 0,\ t \in \mathrm{Clo}_n\mathbf{A},\ \{x_1, \ldots, x_n\} \subseteq X\}.$$

The congruence on $\mathbf{A}$ generated by a set $X \subseteq A^2$ will be denoted $\mathrm{Cg}^{\mathbf{A}}(X)$; we have $\langle a, b\rangle \in \mathrm{Cg}^{\mathbf{A}}(X)$ iff $a = b$ or for some $n \geq 1$ there exist elements $z_0 = a, z_1, \ldots, z_n = b$ of A and $\langle x_1, y_1\rangle, \ldots, \langle x_n, y_n\rangle \in X$ and $p_1, \ldots, p_n \in \mathrm{Pol}_1\mathbf{A}$ such that $\{z_i, z_{i+1}\} = \{p_{i+1}(x_{i+1}), p_{i+1}(y_{i+1})\}$ for all $i < n$.

Given any congruence α on $\mathbf{A}$, we have the quotient algebra $\mathbf{A}/\alpha$ whose elements are the α-equivalence classes u/α, $u \in A$. If $f : \mathbf{A} \to \mathbf{B}$ is a homomorphism, then $\ker f = \{\langle x, y\rangle : f(x) = f(y)\}$ is a congruence on $\mathbf{A}$; conversely, any congruence α on $\mathbf{A}$ is the kernel of the quotient map $p_\alpha : \mathbf{A} \to \mathbf{A}/\alpha$. The **congruence lattice** of an algebra $\mathbf{A}$ is the algebra $\mathbf{Con\,A} = \langle \mathrm{Con}\,\mathbf{A}, \wedge, \vee\rangle$, where $\mathrm{Con}\,\mathbf{A}$ is the set of all congruences on $\mathbf{A}$ and the meet $\alpha \wedge \beta$, and the join $\alpha \vee \beta$, of congruences α and β are, respectively, $\alpha \cap \beta$ and the transitive closure of $\alpha \cup \beta$. The least and greatest congruences on $\mathbf{A}$ are 0_A (the identity relation on A) and 1_A ($= A \times A$). An algebra $\mathbf{A}$ is called **simple** if these two are the only congruences on $\mathbf{A}$, and $|A| > 1$. $\mathbf{A}$ is called **subdirectly irreducible** if it has a congruence $\beta \neq 0_A$ (called the **monolith**) such that the formulas $\alpha \neq 0_A$ and $\beta \leq \alpha$ are equivalent for any congruence α. (We generally use $\leq$ instead of $\subseteq$, and $<$ instead of $\subset$, to denote the order in the congruence lattice.)

To indicate that elements a and b are congruent modulo a congruence α we shall write, variously: $\langle a, b\rangle \in \alpha$, $a\,\alpha\,b$, $a \equiv b \pmod{\alpha}$, or $a \equiv_\alpha b$.

An algebra $\mathbf{A}$ is said to be a **subdirect product** of algebras $\mathbf{A}_t$ ($t \in T$) if $\mathbf{A} \leq \prod\langle \mathbf{A}_t : t \in T\rangle$ ($\mathbf{A}$ is a subalgebra of the direct product of the $\mathbf{A}_t$) and the projection p_t of the product onto $\mathbf{A}_t$ maps $\mathbf{A}$ onto $\mathbf{A}_t$ for every $t \in T$. According to two fundamental results of G. Birkhoff, $\mathbf{A}$ is subdirectly irreducible as defined above if and only if whenever $f : \mathbf{A} \to \prod\langle \mathbf{A}_t : t \in T\rangle$ embeds $\mathbf{A}$ as a subdirect product, then $p_t f$ is an isomorphism of $\mathbf{A}$ with $\mathbf{A}_t$ for one of the projections p_t; and every algebra is isomorphic to a subdirect product of subdirectly irreducible algebras.

By a **graph** we mean a pair $\mathbf{G} = \langle P, E\rangle$ in which P is a nonvoid set (called the set of **vertices** of $\mathbf{G}$) and E is a set of two-element subsets of P (called the set of **edges** of $\mathbf{G}$). This is the definition of graph used in graph theory. For our purposes, we need to construe graphs as models of a first order language. To do this, we correlate to a graph $\mathbf{G} = \langle P, E\rangle$ the structure $\mathbf{G}' = \langle P, B\rangle$ with one binary relation B defined by $\langle x, y\rangle \in B \Leftrightarrow$

$\{x, y\} \in E$. The relation B is symmetric and irreflexive; and it is clear that any symmetric and irreflexive binary relation over P arises in this way from the edge set of a unique graph with vertex set P. Here the distinction between $\mathbf{G}$ and $\mathbf{G}'$ is so slight that we shall often use the word " graph" when we mean a relational structure with one symmetric, irreflexive binary relation.

0.2 Decidability and interpretability

We assume that our readers are familiar with the basic syntactical and semantical notions generally applied to first order languages and their models. (A discussion of all the notions introduced in this section, adequate for our purposes, will be found in Burris, Sankappanavar [1981] Chapter 5; while a more complete development of first order logic can be found, e.g., in J. D. Monk [1976].) The **formulas** of a first order language $\mathsf{L} = (\Phi, \Sigma, \rho)$ are certain well formed expressions consisting of a string of symbols from an alphabet comprised of the operation and relation symbols of L, together with denumerably many variables x_i $(i < \omega)$, two punctuation symbols,) and (, and the symbols $\&, \vee, \neg, \rightarrow, \leftrightarrow, \forall, \exists$ and $\approx$. A formula ψ is a **sentence** of L iff every occurrence of a variable in ψ occurs within the scope of a quantifier $\forall$ or $\exists$—i.e., ψ has no free variables.

Where L is a first order language, $\mathbf{A}$ is a model of L, ψ is a formula of L, and $x_1, \ldots, x_n$ is a list of variables including all the variables that occur free in ψ, and $\bar{a} = \langle a_1, \ldots, a_n \rangle \in A^n$, we use the notation $\mathbf{A} \models \psi(\bar{a})$ to indicate that ψ becomes a true statement when its free variables x_i are replaced by a_i, and its operation and relation symbols are taken to refer to the corresponding operations and relations of $\mathbf{A}$. If ψ is a sentence, we write $\mathbf{A} \models \psi$ to indicate that ψ is a valid (i.e., true) statement about $\mathbf{A}$. The set of all sentences valid in every member of a class $\mathcal{K} \subseteq \mathrm{Mod}(\mathsf{L})$ is called the **theory of** $\mathcal{K}$, and is denoted $\mathrm{Th}(\mathcal{K})$. The class of all L-structures in which every member of a set Γ of sentences is valid is denoted $\mathrm{Mod}(\Gamma)$; and members of $\mathrm{Mod}(\Gamma)$ are called **models of** Γ. A **theory** is a set T of sentences satisfying $\mathsf{T} = \mathrm{ThMod}(\mathsf{T})$; these are just the deductively closed sets of sentences (by the completeness theorem of first order logic).

Intuitively speaking, a decidable theory is one for which there exists an algorithm which will determine, for any given sentence ψ, whether ψ belongs to the theory. Since we wish to prove theorems about decidability of theories in languages that need not be finite, or even countable, we introduce a slightly unusual notion of decidability. Let $\mathcal{K}$ be a class of models of L, and $\mathsf{T} = \mathrm{Th}(\mathcal{K})$ be its theory. First, assume that L is a finite language. Then $\mathcal{K}$ (or T) will be called decidable if T is a recursive set of sentences. (This is the usual notion, for finite languages. We assign Gödel numbers to symbols, and then to sentences, and consider T to be decidable iff the set of Gödel numbers of sentences in T is a recursive set of integers. There are many standard ways to do this, but they all lead to the same result—recursive or non-recursive—for a given theory, since all the standard Gödel numberings are recursively isomorphic.) Now if L is any language, we call $\mathcal{K}$ (or T) **decidable** iff for every finite language $L' \subseteq L$, the set of all sentences in T which are sentences of L' is a recursive set. We call $\mathcal{K}$ (or T) **undecidable** iff it is not decidable. We call $\mathcal{K}$ (or T) **hereditarily undecidable** iff every class $\mathcal{K}'$ of models of L such that $\mathcal{K} \subseteq \mathcal{K}'$ (every theory T' of L such that $T' \subseteq T$) is undecidable.

Since the 1950s, it has become abundantly clear that most theories are undecidable. Undecidability seems to be very closely related to the existence of certain kinds of complexity built into some of the models of a theory. One important objective of the present work is to give a precise form to this intuition, while considering a certain broad class of theories. Nevertheless, we can mention a few theories and classes that are known to be decidable and whose models can be modestly complex. The set of all valid sentences of a language with one unary operation symbol is decidable (A. Ehrenfeucht [1959]). The class of all Abelian groups, as well as every subclass of this class, is decidable (W. Szmielew [1955]). Every class of Boolean algebras is decidable (A. Tarski [1949]).

Undecidability of a theory T is generally proved by the method of **interpretation**; that is, we interpret into T some theory T' that is known to be undecidable. Loosely speaking, this involves finding a way to build a model of T' within a model of T, using first order formulas of the language of T to define the operations and relations of the model of T'. The method requires that every model of T' be built within some model of T, using the same formulas in every case. When this can be done, we say that T' is interpretable into T.

We now define quite precisely a very general notion of interpretation that encompasses all the specific interpretations we shall construct. This precise formulation of the concept of interpretation was first introduced in M. O. Rabin [1965]. (This concept is very different from the concept of interpretation between varieties, i.e., clone homomorphisms, discussed in Hobby, McKenzie [1988], Chapter 9.)

Let L' be a finite language and L be an arbitrary language. We suppose that the operation and relation symbols of L' are $f_1, \dots, f_k$, and $s_1, \dots, s_l$, respectively. A **scheme** for interpreting L' into L is, for some integers $m \geq 1$ and $n \geq 0$, a system

$$\bar{\psi} = \langle \mathrm{Un}, \mathrm{Eq}, F_1, \dots, F_k, S_1, \dots, S_l \rangle$$

of L-formulas such that (where $\bar{y} = \langle y_1, \ldots, y_n \rangle$ and $\bar{x}^i = \langle x^i_1, \ldots, x^i_m \rangle$ for each $i \in \omega$ and all of the variables y_j and x^v_u are distinct)

$$\begin{aligned} \mathrm{Un} &= \mathrm{Un}(\bar{x}^1, \bar{y}), \\ \mathrm{Eq} &= \mathrm{Eq}(\bar{x}^1, \bar{x}^2, \bar{y}), \\ F_a &= F_a(\bar{x}^1, \ldots, \bar{x}^p, \bar{x}^{p+1}, \bar{y}), \text{ where } f_a \text{ is } p\text{-ary}, \\ S_b &= S_b(\bar{x}^1, \ldots, \bar{x}^q, \bar{y}), \text{ where } s_b \text{ is } q\text{-ary}. \end{aligned}$$

(I.e., the free variables of each formula are among those indicated.)

An L-structure $\mathbf{D}$ and an n-tuple $\bar{d}$ from $\mathbf{D}$ **admits** the scheme $\bar{\psi}$ if there is an L$'$-structure

$$\widehat{\mathbf{D}} = \langle \widehat{D}, \hat{f}_1, \ldots, \hat{f}_k, \hat{s}_1, \ldots, \hat{s}_l \rangle$$

and a map ϕ from

$$\mathrm{Un}^{\langle \mathbf{D}, \bar{d} \rangle} = \{\bar{e} \in D^m : \mathbf{D} \models \mathrm{Un}(\bar{e}, \bar{d})\}$$

onto $\widehat{D}$ such that for all $\bar{e}^1, \bar{e}^2, \ldots \in \mathrm{Un}^{\langle \mathbf{D}, \bar{d} \rangle}$,

(1) $\phi(\bar{e}^1) = \phi(\bar{e}^2)$ if and only if $\mathbf{D} \models \mathrm{Eq}(\bar{e}^1, \bar{e}^2, \bar{d})$;

(2) for all $a \le k$, $\mathbf{D} \models F_a(\bar{e}^1, \ldots, \bar{e}^p, \bar{e}^{p+1}, \bar{d})$ if and only if $\hat{f}_a(\phi(\bar{e}^1), \ldots, \phi(\bar{e}^p)) = \phi(\bar{e}^{p+1})$, where f_a is p-ary; and

(3) for all $b \le l$, $\mathbf{D} \models S_b(\bar{e}^1, \ldots, \bar{e}^q, \bar{d})$ if and only if $\langle \phi(\bar{e}^1), \ldots, \phi(\bar{e}^q) \rangle \in \hat{s}_b$, where s_b is q-ary.

Up to isomorphism, $\widehat{\mathbf{D}}$ is determined when $\langle \mathbf{D}, \bar{d} \rangle$ admits $\bar{\psi}$; and we write $\mathbf{D}(\bar{\psi}, \bar{d})$ for the canonical $\widehat{\mathbf{D}}$ with universe consisting of the equivalence classes of the equivalence relation that is $\mathrm{Eq}^{\langle \mathbf{D}, \bar{d} \rangle}$ restricted to $\mathrm{Un}^{\langle \mathbf{D}, \bar{d} \rangle}$.

Definition 0.1

(1) Let $\mathcal{K}'$ be a class of L$'$-structures (for L$'$ as above) and let $\mathcal{K}$ be a class of L-structures. $\mathcal{K}'$ is **interpretable** into $\mathcal{K}$ if there is a scheme $\bar{\psi}$ as above such that for all $\mathbf{A} \in \mathcal{K}'$, $\mathbf{A} \cong \mathbf{D}(\bar{\psi}, \bar{d})$ for some $\mathbf{D} \in \mathcal{K}$ and $\bar{d} \in D$. We say that $\mathcal{K}$ **interprets** $\mathcal{K}'$ via $\bar{\psi}$ if this holds.

(2) A theory T$'$ in L$'$ is **interpretable** into a theory T in L if $\mathcal{K}'$ is interpretable into Mod (T) for some class $\mathcal{K}'$ of L$'$-structures with Th $(\mathcal{K}') =$ T$'$.

THEOREM 0.2 *Let* T$'$ *be a theory in a finite language which is interpretable into a theory* T. *Then,*

(i) *If* T' *is finitely axiomatizable and* T *is decidable, then* T' *is decidable.*

(ii) *If* T' *is hereditarily undecidable (in particular if* T' *is finitely axiomatizable and undecidable) then* T *is hereditarily undecidable.*

PROOF. See Theorem 3.2 of Burris, McKenzie [1981]. □

Remark 0.3 The class $\mathcal{G}_{\mathrm{fin}}$ of all finite graphs is hereditarily undecidable (I.A Lavrov [1963]). This class can easily be interpreted into many others. For example, $\mathcal{G}_{\mathrm{fin}}$ can be interpreted into the class $\mathcal{G}$ of all graphs (trivially), or into the class of all finite L-structures where L is the language with two unary operation symbols. Also, $\mathcal{G}_{\mathrm{fin}}$ can be interpreted into various restricted classes of finite graphs, such as the class of all finite bi-partite graphs. (A graph $\langle P, B\rangle$ is **bi-partite** iff P can be decomposed into the disjoint union of sets S and T so that $B \subseteq (S \times T) \cup (T \times S)$.) The relation "is interpretable into" between classes is a transitive one. In this book, we prove that a class is undecidable by interpreting into it some class of structures of finite type that is known to be hereditarily undecidable, and invoking Theorem 0.2. Usually we use some class of graphs, or the class of Boolean pairs defined in Definition 0.46 below. All of the classes we use interpret the class $\mathcal{G}$ of graphs. Thus, ultimately, $\mathcal{G}$ is interpretable into every variety that is proved in this book to be undecidable.

Now it is well known that $\mathcal{G}$ interprets every class of first order structures of finite type. Thus it turns out that any two varieties of finite type that are proved undecidable in this book interpret one another, as well as interpreting the class $\mathcal{G}$. However, it is not the case that every undecidable variety interprets $\mathcal{G}$. In fact, $\mathcal{G}$ cannot be interpreted into any class of modules over a ring, since such a class is *stable* (in the sense of S. Shelah) and $\mathcal{G}$ is unstable; and there are finite rings for which the class of all modules is undecidable. (See Chapter 14.)

In many cases where we are interpreting some hereditarily undecidable class of graphs directly into a class $\mathcal{K}$, it turns out that the algebras of $\mathcal{K}$ which we use to interpret finite graphs are themselves finite. When we are forced to proceed more indirectly, for instance by interpreting the class of Boolean pairs, then the algebras of $\mathcal{K}$ that are actually involved in the interpretation are necessarily infinite. We shall find it useful to make a distinction between these two possible outcomes, and so we make a definition.

Definition 0.4 Let $\mathcal{K}$ be a class of L-structures for some language L. $\mathcal{K}$ will be called **unstructured** iff the class of all graphs is interpretable into $\mathcal{K}$, and otherwise will be called **structured**. $\mathcal{K}$ will be called **ω-unstructured**

iff the class of all finite graphs is interpretable into the class of finite members of $\mathcal{K}$, and otherwise will be called ω-**structured**.

0.3 Varieties

By a **variety** we mean a class of similar algebras (models of one algebraic language L) closed under the formation of homomorphic images, subalgebras, and direct products. Let L be an algebraic language and $\mathcal{K} \subseteq \operatorname{Mod}(\mathsf{L})$. We use $\boldsymbol{H}(\mathcal{K}), \boldsymbol{S}(\mathcal{K}), \boldsymbol{P}(\mathcal{K})$ to denote, respectively, the class of all homomorphic images of algebras in $\mathcal{K}$, the class of all algebras embeddable into an algebra in $\mathcal{K}$, and the class of all algebras isomorphic to a direct product of a system of algebras in $\mathcal{K}$. We use $\boldsymbol{P}_s(\mathcal{K})$ to denote the class of all algebras isomorphic to a subdirect product of some system of algebras $\mathbf{A}_t \in \mathcal{K}$ $(t \in T)$, and we use $\boldsymbol{P}_{\text{fin}}(\mathcal{K})$ to denote the class of all algebras isomorphic to the product of some finite system of algebras from $\mathcal{K}$. The smallest variety containing $\mathcal{K}$ is $\boldsymbol{HSP}(\mathcal{K})$, also denoted $\boldsymbol{V}(\mathcal{K})$. An **equation** is a sentence of the form $(\forall\bar{x})(s \approx t)$ where s and t are terms of L. By a theorem of G. Birkhoff, the class $\mathcal{K}$ is a variety iff $\mathcal{K} = \operatorname{Mod}(\Gamma)$ for some set Γ of equations; and in general, $\boldsymbol{V}(\mathcal{K})$ is identical with the class of models of all the equations that are valid in $\mathcal{K}$.

Let $\mathcal{V} \subseteq \operatorname{Mod}(\mathsf{L})$ be a variety. The free algebra in $\mathcal{V}$ freely generated by a set of cardinality κ is denoted $\mathbf{F}_{\mathcal{V}}(\kappa)$. An algebra $\mathbf{A}$ is **locally finite** iff every finitely generated subalgebra of $\mathbf{A}$ is finite. The variety $\mathcal{V}$ is locally finite iff every algebra in $\mathcal{V}$ is locally finite, or equivalently, every finitely generated free algebra in $\mathcal{V}$ is finite. The variety $\mathcal{V}$ is said to be **finitely generated** iff $\mathcal{V} = \boldsymbol{V}(\mathbf{A}_1, \ldots, \mathbf{A}_n)$ where n is finite and $\mathbf{A}_1, \ldots, \mathbf{A}_n$ are finite algebras. The join of varieties $\mathcal{V}_1$ and $\mathcal{V}_2$ contained in $\operatorname{Mod}(\mathsf{L})$ is the variety $\boldsymbol{V}(\mathcal{V}_1 \cup \mathcal{V}_2)$, usually denoted by $\mathcal{V}_1 \vee \mathcal{V}_2$. The join of any collection of varieties is the $\boldsymbol{HSP}$-closure of their union.

In dealing with equations, we omit the universal quantifier and write simply $s \approx t$ to denote the equation $(\forall\bar{x})(s \approx t)$. Varieties $\mathcal{V}_1, \ldots, \mathcal{V}_n \subseteq \operatorname{Mod}(\mathsf{L})$ are called **independent** iff there exists an L-term $t(x_1, \ldots, x_n)$ such that $\mathcal{V}_i \models t \approx x_i$ for $i = 1, \ldots, n$. If $\mathcal{V}_1, \ldots, \mathcal{V}_n$ are independent, then every algebra $\mathbf{A}$ in $\mathcal{V} = \mathcal{V}_1 \vee \cdots \vee \mathcal{V}_n$ is isomorphic to a product $\mathbf{A}_1 \times \cdots \times \mathbf{A}_n$ with $\mathbf{A}_1 \in \mathcal{V}_1, \ldots, \mathbf{A}_n \in \mathcal{V}_n$ and the algebras $\mathbf{A}_i$ are determined up to isomorphism. In this case, we write $\mathcal{V}_1 \otimes \cdots \otimes \mathcal{V}_n$ for the join variety $\mathcal{V} = \mathcal{V}_1 \vee \cdots \vee \mathcal{V}_n$, and say that $\mathcal{V}$ is the **product** of its subvarieties $\mathcal{V}_1, \ldots, \mathcal{V}_n$. We shall need the following result in Chapter 13.

THEOREM 0.5 *Let $\mathcal{V}$ be a locally finite variety and $\mathcal{V}_1, \ldots, \mathcal{V}_n$ be subvarieties of $\mathcal{V}$ such that $\mathcal{V} = \mathcal{V}_1 \vee \cdots \vee \mathcal{V}_n$. The following are equivalent.*

(i) *Every subdirect product* $\mathbf{A} \leq \mathbf{A}_1 \times \cdots \times \mathbf{A}_n$ *with* $\mathbf{A}_i \in \mathcal{V}_i$ $(1 \leq i \leq n)$ *is equal to the direct product.*

(ii) *Every subdirect product* $\mathbf{A} \leq \mathbf{A}_1 \times \cdots \times \mathbf{A}_n$ *with* $\mathbf{A}_i \in \mathcal{V}_i$ $(1 \leq i \leq n)$ *and* $\mathbf{A}$ *finite is equal to the direct product.*

(iii) *If* $\mathbf{A} = \mathbf{F}_{\mathcal{V}}(n)$ *and* $\mathbf{A}_i = \mathbf{F}_{\mathcal{V}_i}(n)$ *then* $\mathbf{A} \cong \mathbf{A}_1 \times \cdots \times \mathbf{A}_n$.

(iv) $\mathcal{V} = \mathcal{V}_1 \otimes \cdots \otimes \mathcal{V}_n$.

PROOF. Refer to §0.1 for the concept of a subdirect product. To see that (iv) $\Rightarrow$ (i), let $t(x_1, \ldots, x_n)$ be a term such that $\mathcal{V}_i \models t \approx x_i$. Suppose that $\mathbf{A}$ is a subdirect product of $\mathbf{A}_i \in \mathcal{V}_i$. Given $\langle a_1, \ldots, a_n \rangle \in A_1 \times \cdots \times A_n$, choose $u_i \in A$ such that $p_i(u_i) = a_i$ for $1 \leq i \leq n$. Then the element $u = t^{\mathbf{A}}(u_1, \ldots, u_n) \in A$ can be none other than $\langle a_1, \ldots, a_n \rangle$, since for each i,

$$p_i(u) = t^{\mathbf{A}_i}(p_i(u_1), \ldots, p_i(u_n)) = p_i(u_i) = a_i.$$

The implications (i) $\Rightarrow$ (ii) $\Rightarrow$ (iii) are obvious. To see that (iii) $\Rightarrow$ (iv), let $\mathbf{A}_i = \mathbf{F}_{\mathcal{V}_i}(x_1, \ldots, x_n)$ for $1 \leq i \leq n$, and let $\mathbf{A} = \mathbf{A}_1 \times \cdots \times \mathbf{A}_n$. For $j = 1, \ldots, n$ let $y_j \in A$ be $\langle x_j, x_j, \ldots, x_j \rangle$, and let $\mathbf{B}$ be the subalgebra of $\mathbf{A}$ generated by $\{y_1, \ldots, y_n\}$. If s and t are terms of L such that $s^{\mathbf{A}}(\bar{y}) = t^{\mathbf{A}}(\bar{y})$ (where $\bar{y} = \langle y_1, \ldots, y_n \rangle$) then $\mathcal{V}_i \models s \approx t$ for $i \in \{1, \ldots, n\}$, and so $\mathcal{V} \models s \approx t$. Thus $\mathbf{B}$ is freely generated over $\mathcal{V}$ by $y_1, \ldots, y_n$. Then by (iii), $\mathbf{B} \cong \mathbf{A}$, and since $\mathbf{B} \leq \mathbf{A}$ and $\mathbf{A}$ is finite, it follows that $\mathbf{B} = \mathbf{A}$. There must then exist some term t such that $t^{\mathbf{A}}(\bar{y}) = \langle x_1, x_2, \ldots, x_n \rangle$. Projecting to the ith coordinate, we find that $t^{\mathbf{A}_i}(\bar{x}) = x_i$, which is equivalent to $\mathcal{V}_i \models t \approx x_i$, since $\mathbf{A}_i$ is the free algebra in $\mathcal{V}_i$. □

0.4 Abelian and solvable algebras

The special classes of universal algebras called Abelian, solvable, affine, strongly solvable, and strongly Abelian will play important roles in this book. Recall that $\mathrm{Clo}_k \mathbf{A}$ denotes the set of all k-ary term operations of an algebra $\mathbf{A}$, and $\mathrm{Con}\,\mathbf{A}$ denotes the set of all congruence relations of $\mathbf{A}$.

Definition 0.6 Let $\alpha, \beta, \gamma \in \mathrm{Con}\,\mathbf{A}$. We write $\mathrm{C}(\alpha, \beta; \gamma)$, and say that α **centralizes** β **modulo** γ, provided that the following condition holds:

For every $n \geq 1$, for every $f \in \mathrm{Clo}_{n+1}\mathbf{A}$, and for all $\langle a, b \rangle \in \alpha$ and $\langle c_1, d_1 \rangle, \ldots, \langle c_n, d_n \rangle \in \beta$ we have

$$f(a, \bar{c}) \equiv_\gamma f(a, \bar{d}) \leftrightarrow f(b, \bar{c}) \equiv_\gamma f(b, \bar{d}).$$

This concept of a congruence centralizing a congruence modulo a third congruence has the following properties: $C(\alpha, \beta; \alpha \wedge \beta)$ always holds; if $C(\alpha_t, \beta; \gamma)$ for all $t \in T$ then $C(\bigvee_{t \in T} \alpha_t, \beta; \gamma)$; if $C(\alpha, \beta; \gamma_t)$ for all $t \in T$ then $C(\alpha, \beta; \bigwedge_{t \in T} \gamma_t)$. From this it follows that for $\alpha, \beta \in \operatorname{Con} \mathbf{A}$, there exists a largest congruence δ satisfying $C(\delta, \alpha; \beta)$, called the **centralizer of α modulo β**, and denoted $(\beta : \alpha)$; and there exists a smallest congruence λ satisfying $C(\alpha, \beta; \lambda)$, called the **commutator** of α and β, and denoted $[\alpha, \beta]$. The centralizer of 1_A modulo 0_A (with respect to the algebra $\mathbf{A}$) is called the center of $\mathbf{A}$. This congruence of $\mathbf{A}$ can be defined more directly.

Definition 0.7 Let $\mathbf{A}$ be any algebra. The **center** of $\mathbf{A}$ is the binary relation $Z(\mathbf{A})$ defined by $\langle x, y \rangle \in Z(\mathbf{A}) \longleftrightarrow$

for all $n \geq 1$, for all $f \in \operatorname{Clo}_{n+1} \mathbf{A}$, and for all $c_1, d_1, \ldots, c_n, d_n \in A$

$$f(x, \bar{c}) = f(x, \bar{d}) \leftrightarrow f(y, \bar{c}) = f(y, \bar{d}).$$

It is a congruence on $\mathbf{A}$. The algebra $\mathbf{A}$ is called **Abelian** iff $Z(\mathbf{A}) = 1_A$, and called **centerless** iff $Z(\mathbf{A}) = 0_A$.

This concept of an Abelian algebra is considerably more inclusive than the concept used in Burris, McKenzie [1981]. What was called in that paper an Abelian algebra will here be called an affine algebra.

Definition 0.8 Algebras $\mathbf{A}$ and $\mathbf{B}$ are said to be **polynomially equivalent** if they have the same universe and precisely the same polynomial operations, i.e., if $\operatorname{Pol} \mathbf{A} = \operatorname{Pol} \mathbf{B}$. An algebra $\mathbf{A}$ is called **affine** iff $\mathbf{A}$ is polynomially equivalent with an algebra $\mathbf{M}$ that is a module over a ring.

THEOREM 0.9 *An algebra $\mathbf{A}$ is affine if and only if it satisfies one of these conditions (which are equivalent).*

(i) *$\mathbf{A}$ is Abelian and possesses a polynomial operation $p(x, y, z)$ obeying the equations $p(x, y, y) \approx x$ and $p(x, x, y) \approx y$.*

(ii) *$\mathbf{A}$ possesses a term operation $p(x, y, z)$ obeying the above equations and such that for every basic operation f of $\mathbf{A}$, the equation*

$$p(f(\bar{x}), f(\bar{y}), f(\bar{z})) \approx f(p(x_0, y_0, z_0), \ldots, p(x_{n-1}, y_{n-1}, z_{n-1}))$$

is valid in $\mathbf{A}$ (if f is n-ary).

PROOF. See Burris, Sankappanavar [1981], Chapter 2 §13 or Chapter 3 of Hobby, McKenzie [1988]. It is a good exercise to show that if there exists a polynomial operation satisfying either (i) or (ii) then it can be taken to be a term operation. □

Definition 0.10 Let **A** be any algebra, $\alpha, \beta, \gamma, \delta$ be congruences on **A**.

(1) If $\alpha \leq \beta$, we say that β is **Abelian over** α iff $C(\beta, \beta; \alpha)$ (see Definition 0.6).

(2) If $\delta \leq \gamma$, we say that γ is **solvable over** δ iff there exists a finite chain of congruences $\alpha_0 = \delta \leq \alpha_1 \leq \cdots \leq \alpha_n = \gamma$ with α_{i+1} Abelian over α_i for all $i < n$.

(3) If $\delta \leq \gamma$, we say that γ is **locally solvable over** δ iff for every finitely generated subalgebra $\mathbf{B} \leq \mathbf{A}$ and for the restricted congruences $\delta|_B$ and $\gamma|_B$ of **B**, we have that $\gamma|_B$ is solvable over $\delta|_B$.

(4) We say that **A** is **solvable** iff 1_A is solvable over 0_A; **locally solvable** iff 1_A is locally solvable over 0_A.

THEOREM 0.11 *The relation "$\alpha \vee \beta$ is locally solvable over $\alpha \wedge \beta$" defines a congruence on the congruence lattice of any locally finite algebra* **A**. *If $\alpha \leq \beta$ are congruences of a locally finite algebra* **A**, *then β is locally solvable over α iff there fails to exist a pair $\langle u, v\rangle \in \beta - \alpha$ and and a binary polynomial operation $p(x, y)$ such that $p(u, u) = p(u, v) = p(v, u) = u$ and $p(v, v) = v$.*

PROOF. See Hobby, McKenzie [1988], Chapter 7 (especially Theorem 7.2, Lemma 7.4). □

Definition 0.12 Let $\alpha \leq \beta$ be congruences of an algebra **A**. We say that β is **strongly Abelian** over α iff for all $n \geq 1$, for all $f \in \mathrm{Clo}_{n+1}\mathbf{A}$, and for all $a \equiv b \equiv c \pmod{\beta}$ and $\langle u_1, v_1\rangle, \ldots, \langle u_n, v_n\rangle \in \beta$ we have

$$f(a, \bar{u}) \equiv_\alpha f(b, \bar{v}) \rightarrow f(c, \bar{u}) \equiv_\alpha f(c, \bar{v}).$$

We say that **A** is **strongly Abelian** iff 1_A is strongly Abelian over 0_A.

Definition 0.13 Let $\delta \leq \gamma$ be congruences of an algebra **A**.

(1) We say that γ is **strongly solvable** over δ iff there exists a finite chain of congruences $\alpha_0 = \delta \leq \alpha_1 \leq \cdots \leq \alpha_n = \gamma$ such that α_{i+1} is strongly Abelian over α_i for all $i < n$.

(2) We say that γ is **locally strongly solvable** over δ iff for every finitely generated subalgebra $\mathbf{B} \leq \mathbf{A}$ and for the restricted congruences $\delta|_B$ and $\gamma|_B$ of the algebra **B**, we have that $\gamma|_B$ is strongly solvable over $\delta|_B$.

(3) The algebra **A** is said to be **strongly solvable** iff 1_A is strongly solvable over 0_A; and is said to be **locally strongly solvable** iff 1_A is locally strongly solvable over 0_A.

The next two theorems are from Hobby, McKenzie [1988] (Lemma 7.4, Corollary 7.6).

THEOREM 0.14 *The relation "$\alpha \vee \beta$ is locally strongly solvable over $\alpha \wedge \beta$" defines a congruence on the congruence lattice of any locally finite algebra* **A**. *If $\alpha \le \beta$ are congruences of a locally finite algebra* **A**, *then β is locally strongly solvable over α iff there fails to exist a pair $\langle u, v\rangle \in \beta - \alpha$ and a binary polynomial operation $p(x, y)$ such that $p(u, v) = p(v, u) = u$ and $p(v, v) = v$.*

THEOREM 0.15 *For every locally finite variety $\mathcal{V}$, the class of all locally solvable algebras in $\mathcal{V}$, and the class of all locally strongly solvable algebras in $\mathcal{V}$, are varieties.*

Example 0.16 We offer several examples for the concepts of this section. (It is not difficult to work out the details.)

1. If $K \subseteq L$ are normal subgroups of a group **G** and $\kappa \le \lambda$ are the corresponding congruences of **G**, then λ is Abelian over κ iff $[L, L] \subseteq K$. (Moreover, $\mathrm{C}(\lambda, \kappa; 0_G)$ iff $[L, K] = \{e\}$.) Thus **G** is solvable in the sense defined above iff **G** is a solvable group.

2. If $J \subseteq K$ are ideals of a ring **R** and $\eta \le \kappa$ are the corresponding congruences, then κ is Abelian over η iff $K \cdot K \subseteq J$. **R** is Abelian in our sense iff **R** is a zero ring (with trivial multiplication).

3. Every module is Abelian in our sense. A module, ring, or group with more than one element cannot be strongly solvable. For an example of a strongly Abelian algebra, consider any algebra that has no operation depending on more than one variable.

We will need the following facts in Part II.

THEOREM 0.17 *Let* **A** *be an algebra.*

(i) *If $\alpha \le \beta$ in* Con **A** *and β is strongly Abelian over α, then β is Abelian over α.*

(ii) **A** *is strongly Abelian iff for all $n \ge 1$ and all $p \in \mathrm{Pol}_n \mathbf{A}$, there are equivalence relations $E_1, \ldots, E_n$ on A such that for $\bar{a}, \bar{b} \in A^n$, $p(\bar{a}) = p(\bar{b})$ if and only if $\langle a_i, b_i\rangle \in E_i$ for all $i < n$.*

(iii) *If* $\mathbf{A}$ *is strongly Abelian and is finite, then any polynomial operation of* $\mathbf{A}$ *depends essentially on at most* $\log_2(|A|)$ *of its variables.*

(iv) *Every locally finite variety, all of whose members are strongly Abelian, is finitely generated.*

(v) *Every locally finite variety of finite type, all of whose members are strongly Abelian, is finitely axiomatizable.*

PROOF. The first two assertions follow quite directly from the definitions. The third follows immediately from the second.

To prove (iv), let $\mathbf{A} = \mathbf{F}_{\mathcal{V}}(x, y)$ and $k = |A|$. Let $t(x_1, \ldots, x_n)$ be any term of $\mathcal{V}$. If $n \geq k$ then some two of the elements

$$t^{\mathbf{A}}(y, \ldots, y), t^{\mathbf{A}}(x, y, \ldots, y), \ldots, t^{\mathbf{A}}(x, \ldots, x)$$

are equal. If these two equal elements have an x and a y in the ith place, then the strongly Abelian property implies that $t^{\mathbf{B}}$ is independent of its ith variable for every $\mathbf{B} \in \mathcal{V}$. Thus there does not exist a term that in $\mathcal{V}$ depends on k or more variables. From this, it easily follows that any equation that holds in the algebra $\mathbf{F}_{\mathcal{V}}(2k)$ holds in every algebra of $\mathcal{V}$. Thus $\mathcal{V}$ is generated by its free algebra $\mathbf{F}_{\mathcal{V}}(2k)$, which is a finite algebra.

To prove (v), let k be as above. Let $t_0, \ldots, t_n$ be a finite set of terms in the variables $x_0, \ldots, x_{k-1}$ so that every term $t(x_0, \ldots, x_{k-1})$ is equal in $\mathcal{V}$ to some $t_i(x_0, \ldots, x_{k-1})$. Let Γ be a finite set of equations asserting that the fundamental operations of $\mathcal{V}$ depend only on the variables upon which they actually depend in $\mathcal{V}$. For each $i, i_0, \ldots, i_{k-1} \leq n$, there is some $j \leq n$ and $j_0, \ldots, j_{k-1} < k^2$ such that the equation

$$\begin{aligned} t_i(t_0(x_0, \ldots, x_{k-1}), t_1(x_k, \ldots, x_{2k-1}), \quad \ldots \quad , t_{k-1}(x_{k^2-k}, \ldots, x_{k^2-1})) \approx \\ \approx \quad t_j(x_{j_0}, \ldots, x_{j_{k-1}}) \end{aligned}$$

is valid in $\mathcal{V}$. Let Σ be the (finite) collection of all equations of this type that are valid in $\mathcal{V}$. It is well-known that there is a finite set Λ of equations valid in $\mathcal{V}$ such that every equation of k variables valid in $\mathcal{V}$ is derivable from Λ. Now the set $\Gamma \cup \Sigma \cup \Lambda$ axiomatizes $\mathcal{V}$. The proof of this fact is left for the reader to discover. □

0.5 Special kinds of varieties

We introduce the families of Abelian varieties, strongly Abelian varieties, affine varieties, and discriminator varieties. The last two are contained in the broad family of Maltsev varieties.

Two equivalence relations α and β on a set A are said to **permute** iff whenever $a\,\alpha\,b\,\beta\,c$ there exists some element d such that $a\,\beta\,d\,\alpha c$. A. I. Maltsev [1954] proved that a variety $\mathcal{V}$ has the property that every two congruences on any algebra in $\mathcal{V}$ permute iff there exists a term $t(x,y,z)$ in the language of $\mathcal{V}$ for which the equations $t(x,y,y) \approx x$ and $t(x,x,y) \approx y$ are valid in $\mathcal{V}$. When such a term exists, we say that $\mathcal{V}$ is **Maltsev**. An operation on a set A that obeys these two equations on A is called a Maltsev operation; also, an algebra having a term operation that obeys these equations is called a Maltsev algebra. A useful corollary of Maltsev's result is that a variety $\mathcal{V}$ is Maltsev if and only if the free algebra $\mathbf{F}_{\mathcal{V}}(3)$ has permuting congruences.

Definition 0.18 If (P) is any one of the properties "Abelian", "affine", "strongly Abelian", "locally solvable", "locally strongly solvable" defined in the last section, we say that a variety is (P) iff every algebra in the variety is (P).

The basic results concerning affine algebras and affine varieties are fully detailed in Freese, McKenzie [1987]. We describe these results here without giving proofs. Note that if $\mathcal{V}$ is affine, then, since the free algebra on three generators in $\mathcal{V}$ has a Maltsev term operation, it follows that $\mathcal{V}$ is Maltsev.

A variety $\mathcal{V}$ is said to be **congruence-modular** (or **congruence-distributive**) iff the congruence lattice of each algebra in $\mathcal{V}$ is a modular (or distributive) lattice.

THEOREM 0.19 *A variety $\mathcal{V}$ is affine if and only if it is congruence-modular and Abelian. If $\mathcal{V}$ is affine then it is in fact Maltsev; and there exists a term $t(x,y,z)$ in the language of $\mathcal{V}$ and a ring $\mathbf{R}$ with unit such that every algebra in $\mathcal{V}$ is polynomially equivalent with a unitary left $\mathbf{R}$-module in which $x - y + z = t(x,y,z)$, and every unitary left $\mathbf{R}$-module is polynomially equivalent with an algebra in $\mathcal{V}$.*

Concerning Maltsev algebras in general, we shall need the following facts.

THEOREM 0.20 *Suppose that the algebra $\mathbf{A}$ has a Maltsev polynomial operation $p(x,y,z)$.*

(i) *Every subuniverse of $\mathbf{A}^2$ containing 0_A is a congruence on $\mathbf{A}$.*

(ii) *Let $X \cup \{\langle a,b\rangle\} \subseteq A^2$. Then $\langle a,b\rangle$ belongs to the least congruence including X iff there exist finitely many pairs $\langle x_0,y_0\rangle,\ldots,\langle x_{n-1},y_{n-1}\rangle \in X$ and a polynomial operation $f \in \operatorname{Pol}_n \mathbf{A}$ such that $f(x_0\ldots x_{n-1}) = a$ and $f(y_0\ldots y_{n-1}) = b$.*

PROOF. Observe that a subset $S \subseteq A^2$ is a subuniverse and includes 0_A iff S is closed under each polynomial operation of **A** acting coordinatewise in A^2. Then (ii) follows from (i), since the condition stated in (ii) is obviously necessary and sufficient in order that $\langle a, b\rangle$ belong to the least subset of A^2 that contains X and is closed under all polynomial operations of **A** acting coordinatewise in A^2. To prove (i), let $r \subseteq A^2$ be a subuniverse with $\langle x, x\rangle \in r$ for all $x \in A$. Then r will be a congruence iff it is a symmetric and transitive relation on A. To prove symmetry, let $\langle c, d\rangle \in r$. Note that

$$\langle d, c\rangle = p^{\mathbf{A}^2}(\langle d, d\rangle, \langle c, d\rangle, \langle c, c\rangle).$$

To prove transitivity, assume that $\langle c, d\rangle, \langle d, e\rangle \in r$, and observe that

$$\langle c, e\rangle = p^{\mathbf{A}^2}(\langle c, d\rangle, \langle d, d\rangle, \langle d, e\rangle).$$

□

The concept of a discriminator variety is in many respects the polar opposite of that of an Abelian variety. On any set U we can define an operation $t_U(x, y, z)$ by stipulating that $t_U(x, y, z)$ is z if $x = y$, and is x if $x \neq y$. This operation t_U is called the **ternary discriminator** on U.

Definition 0.21 A variety $\mathcal{V}$ is called a **discriminator variety** iff there exists a term $t(x, y, z)$ in the language of $\mathcal{V}$ such that $\mathcal{V} = \boldsymbol{V}(\mathcal{S})$ where $\mathcal{S}$ is the class of all $\mathbf{A} \in \mathcal{V}$ such that $t^{\mathbf{A}} = t_A$ (i.e., the term t defines the discriminator on the universe of **A**). Such a term t is called a **discriminator term** for $\mathcal{V}$.

There is a very nice structure theory for discriminator varieties, the details of which can be found in Burris, Sankappanavar [1981]. Several important facts about these varieties are given in Theorem 0.22 below. An algebra **A** is called **hereditarily simple** iff $|A| > 1$ and every subalgebra $\mathbf{B} \leq \mathbf{A}$ with more than one element is simple. A variety $\mathcal{V}$ is called **arithmetical** iff $\mathcal{V}$ is congruence-distributive and Maltsev.

THEOREM 0.22 *Let $\mathcal{V}$ be a discriminator variety with discriminator term t. Then $\mathcal{V}$ is an arithmetical variety. The equations*

$$t(x, y, y) \approx t(x, y, x) \approx t(y, y, x) \approx x$$

*are valid in $\mathcal{V}$. Every algebra in $\mathcal{V}$ is centerless; and every finite algebra in $\mathcal{V}$ is isomorphic to a direct product of simple algebras. The following are equivalent for every $\mathbf{A} \in \mathcal{V}$: **A** is subdirectly irreducible; **A** is hereditarily simple; $t^{\mathbf{A}} = t_A$.*

PROOF. See Burris, Sankappanavar [1981], Chapter 4, §§9–10. □

A very natural example of a discriminator variety is the variety of Boolean algebras. The only subdirectly irreducible Boolean algebra (up to isomorphism) is the two-element algebra $\langle \{0,1\}, \wedge, \vee, ^- \rangle$. The term

$$(x \wedge y^-) \vee (x \wedge y \wedge z) \vee (x^- \wedge y^- \wedge z)$$

is a discriminator term for Boolean algebras.

0.6 Tame congruence theory

In this section we present the fragment of tame congruence theory that is required in this book. The reader may wish to refer to Hobby, McKenzie [1988] for more details.

For $\mathbf{A}$ an algebra and U a nonvoid subset of A, $(\mathrm{Pol}\,\mathbf{A})|_U$ denotes the set of all $f|_U$ where $f \in \mathrm{Pol}\,\mathbf{A}$ and U is closed under f. (We write $f|_U$ for the restriction of f to U^n. Recall that the clone $\mathrm{Pol}\,\mathbf{A}$ of polynomial operations of $\mathbf{A}$ was defined in §0.1.) The non-indexed algebra $\mathbf{A}|_U = \langle U\ ,\ (\mathrm{Pol}\,\mathbf{A})|_U \rangle$ is called the **algebra induced by A on** U. Note that the set $(\mathrm{Pol}\,\mathbf{A})|_U$ of basic operations of this algebra is actually a polynomial clone on the set U.

For non-indexed algebras $\langle A, F \rangle$ and $\langle B, G \rangle$, we call a map $f : A \to B$ a homomorphism if the sets F and G can be indexed as $F = \{f_i : i \in I\}$ and $G = \{g_i : i \in I\}$ in such a way that the indexed algebras $\langle A, f_i(i \in I) \rangle$ and $\langle B, g_i(i \in I) \rangle$ are of the same type and f is a homomorphism between them. The congruences, subuniverses, and polynomial operations of a non-indexed algebra $\langle A, F \rangle$ are the same as those of any indexed algebra $\langle A, f_i(i \in I) \rangle$ satisfying $F = \{f_i : i \in I\}$.

Definition 0.23

(1) Let μ, ν be two binary relations on a set A and let $f : A \to A$. We say that f **collapses** μ into ν and write $f(\mu) \subseteq \nu$ if $\langle f(a), f(b) \rangle \in \nu$ for all $\langle a, b \rangle \in \mu$.

(2) A function $e : A \to A$ is **idempotent** if $e(x) = e(e(x))$ for all $x \in A$. We let

$$\mathrm{E}(\mathbf{A}) = \{e \in \mathrm{Pol}_1 \mathbf{A}\ :\ e \text{ is idempotent}\}.$$

We leave it to the reader to prove that if A is finite and $f : A \to A$, then for some $n > 0$, f^n is idempotent.

Definition 0.24 Let $\mathbf{A}$ be a finite algebra and let $\alpha < \beta \in \operatorname{Con} \mathbf{A}$. Let

$$U_{\mathbf{A}}(\alpha,\beta) = \{f(A) : f \in \operatorname{Pol}_1 \mathbf{A} \text{ and } f(\beta) \not\subseteq \alpha\}$$

and $M_{\mathbf{A}}(\alpha,\beta)$ be the set of all minimal members of $U_{\mathbf{A}}(\alpha,\beta)$ relative to the ordering of inclusion. A member of $M_{\mathbf{A}}(\alpha,\beta)$ is called an **$\langle\alpha,\beta\rangle$-minimal set** of $\mathbf{A}$.

By a **congruence quotient** of an algebra $\mathbf{A}$ we mean a pair $\langle\alpha,\beta\rangle$ of congruences of $\mathbf{A}$ such that $\alpha < \beta$. When $\alpha \leq \beta$, we use $I[\alpha,\beta]$ to denote the interval $\{\delta \in \operatorname{Con} \mathbf{A} : \alpha \leq \delta \leq \beta\}$. A congruence quotient $\langle\alpha,\beta\rangle$ of $\mathbf{A}$ is called a **prime quotient** iff β *covers* α, i.e. $|I[\alpha,\beta]| = 2$. The relation of covering between two elements of $\operatorname{Con} \mathbf{A}$ is written $\alpha \prec \beta$.

For $\mathbf{A}$ an algebra and $U, V \subseteq A$, we say that U and V are **polynomially isomorphic**, and write $U \simeq V$, iff there are $f, g \in \operatorname{Pol}_1 \mathbf{A}$ with

$$f(U) = V\,,\; g(V) = U\,,\; fg|_V = id|_V \text{ and } gf|_U = id|_U.$$

If $f \in \operatorname{Pol}_1 \mathbf{A}$ then we write $f : U \simeq V$, and say that f is a **polynomial isomorphism** of U onto V, iff there exists $g \in \operatorname{Pol}_1 \mathbf{A}$ so that the above formulas are valid.

When $\alpha \prec \beta$ then much can be said about the $\langle\alpha,\beta\rangle$-minimal sets (Definition 0.24). A fundamental theorem is the following one.

THEOREM 0.25 *Let $\mathbf{A}$ be a finite algebra and let $\langle\alpha,\beta\rangle$ be a prime congruence quotient of $\mathbf{A}$.*

(i) *For all $U, V \in M_{\mathbf{A}}(\alpha,\beta)$, $U \simeq V$ and so $\mathbf{A}|_U \cong \mathbf{A}|_V$.*

(ii) *For all $U \in M_{\mathbf{A}}(\alpha,\beta)$, there is some $e \in \mathrm{E}(\mathbf{A})$ such that $e(A) = U$.*

(iii) *For all $U \in M_{\mathbf{A}}(\alpha,\beta)$ and $f \in \operatorname{Pol}_1 \mathbf{A}$, if $f(\beta|_U) \not\subseteq \alpha$ then $f(U) \in M_{\mathbf{A}}(\alpha,\beta)$ and $f : U \simeq f(U)$.*

(iv) *If $\langle x, y\rangle \in \beta - \alpha$ and $U \in M_{\mathbf{A}}(\alpha,\beta)$, then for some $f \in \operatorname{Pol}_1 \mathbf{A}$ we have $f(A) = U$ and $\langle f(x), f(y)\rangle \in \beta - \alpha$.*

(v) *For all $f \in \operatorname{Pol}_1 \mathbf{A}$, if $f(\beta) \not\subseteq \alpha$ then for some $U \in M_{\mathbf{A}}(\alpha,\beta)$, $f : U \simeq f(U)$.*

PROOF. See the proof of Theorem 2.8 in Hobby, McKenzie [1988]. □

A more general version of the above theorem is given in Hobby, McKenzie, where the word "prime" is replaced by "tame". The notion of tameness is rather technical and is not needed in this book. For us, it is sufficient to note that every prime congruence quotient of a finite algebra is tame.

Definition 0.26 Let $\mathbf{A}$ be a finite algebra and let $\alpha \prec \beta$ in $\operatorname{Con}\mathbf{A}$. Then we say that $\mathbf{A}$ is **$\langle\alpha,\beta\rangle$-minimal** if $A \in M_{\mathbf{A}}(\alpha,\beta)$. Also, $\mathbf{A}$ is called **minimal** if $\mathbf{A}$ is $\langle 0_A, 1_A\rangle$-minimal.

LEMMA 0.27 *Let $\langle\alpha,\beta\rangle$ be a prime congruence quotient of the finite algebra $\mathbf{A}$.*

(i) *$\mathbf{A}$ is $\langle\alpha,\beta\rangle$-minimal if and only if for all $f \in \operatorname{Pol}_1\mathbf{A}$, either f is a permutation of A or $f(\beta) \subseteq \alpha$.*

(ii) *If $U \in M_{\mathbf{A}}(\alpha,\beta)$ then the induced algebra $\mathbf{A}|_U$ is $\langle\alpha|_U, \beta|_U\rangle$-minimal.*

PROOF. See the proof of Lemma 2.13 in Hobby, McKenzie. □

The following theorem is due to P. P. Palfy and serves to characterize the finite minimal algebras.

THEOREM 0.28 *Every minimal algebra of at least three elements, that has a polynomial operation that depends on more than one variable, is polynomially equivalent with a vector space over a finite field.*

PROOF. See the proof of Theorem 4.7 in Hobby, McKenzie. □

Every two-element algebra is a minimal algebra. Up to polynomial equivalence, there are altogether five different kinds of minimal algebras. We list them all in the following definition.

Definition 0.29 Let $\mathbf{A}$ be minimal.

(1) $\mathbf{A}$ is of type **1** if $\mathbf{A}$ is essentially unary.

(2) $\mathbf{A}$ is of type **2** if $\mathbf{A}$ is polynomially equivalent with a vector space.

(3) $\mathbf{A}$ is of type **3** if $\mathbf{A}$ is polynomially equivalent with a two-element Boolean algebra.

(4) $\mathbf{A}$ is of type **4** if $\mathbf{A}$ is polynomially equivalent with a two-element lattice.

(5) $\mathbf{A}$ is of type **5** if $\mathbf{A}$ is polynomially equivalent with a two-element semilattice.

THEOREM 0.30 *A finite algebra is minimal if and only if it is of one of the five types listed above.*

PROOF. See Corollary 4.11 in Hobby, McKenzie. □

It is also possible to classify algebras that are minimal relative to a prime quotient, again into five different types.

Definition 0.31 Let **A** and **C** be finite algebras and suppose that **C** is $\langle\delta,\theta\rangle$-minimal, $\delta \prec \theta \in \mathrm{Con}\,\mathbf{C}$, and that $\alpha \prec \beta \in \mathrm{Con}\,\mathbf{A}$. By a $\langle\delta,\theta\rangle$**-trace** in **C** we mean any set $N \subseteq C$ of the form x/θ such that $x/\theta \neq x/\delta$. By an $\langle\alpha,\beta\rangle$-trace in **A** we mean any set $N \subseteq A$ such that for some $U \in M_{\mathbf{A}}(\alpha,\beta)$, $N \subseteq U$ and N is an $\langle\alpha|_U,\beta|_U\rangle$-trace of the $\langle\alpha|_U,\beta|_U\rangle$-minimal algebra $\mathbf{A}|_U$. (Thus, $N = (x/\beta)\cap U$ for some $x \in U$ such that $(x/\alpha)\cap U \neq (x/\beta)\cap U$.) The **body** and the **tail** of **C** with respect to $\langle\delta,\theta\rangle$ are defined by

$$\text{body} = \bigcup\{\langle\delta,\theta\rangle\text{-traces}\},$$

$$\text{tail} = C - \text{body}.$$

The body and tail of an $\langle\alpha,\beta\rangle$-minimal set U are defined in a similar way.

LEMMA 0.32 *Let $\alpha \prec \beta$ in* **Con A** *where* **A** *is a finite algebra.*

(i) *Let N be an $\langle\alpha,\beta\rangle$-trace and let $g \in \mathrm{Pol}_1\mathbf{A}$ such that $g(\beta|_N) \nsubseteq \alpha$. Then $g(N)$ is an $\langle\alpha,\beta\rangle$-trace.*

(ii) *Let*
$$\rho = \alpha \cup \{N^2 : N \text{ is an } \langle\alpha,\beta\rangle \text{ trace}\}.$$
Then $\beta =$ the transitive closure of ρ in A.

(iii) *Let N be an $\langle\alpha,\beta\rangle$-trace contained in the $\langle\alpha,\beta\rangle$-minimal set U. Then $\mathbf{A}|_N = (\mathbf{A}|_U)|_N$ and the algebra $(\mathbf{A}|_N)/(\alpha|_N)$ is minimal. Furthermore, since $\langle\alpha,\beta\rangle$ is prime, the just mentioned algebra is simple.*

PROOF. Consult Lemmas 2.4 and 2.17 and the discussion following Corollary 4.11 in Hobby, McKenzie. □

Another way to state (ii) of the above lemma is as follows. If $\alpha \prec \beta$ in **Con A** and $\langle a,b\rangle \in \beta-\alpha$, then modulo α we can connect a to b via a series of intersecting $\langle\alpha,\beta\rangle$-traces, i.e., for some n and $\langle\alpha,\beta\rangle$-traces $N_1, N_2, \ldots, N_n$ of **A**, we have

$$(a/\alpha)\cap N_1 \neq \emptyset \neq (b/\alpha)\cap N_n$$

and

$$(N_i/\alpha)\cap(N_{i+1}/\alpha) \neq \emptyset$$

for all $i < n$. This connectivity property will be used quite a bit in Part II.

Since $(\mathbf{A}|_N)/(\alpha|_N)$ is a minimal algebra for any finite algebra $\mathbf{A}$, prime quotient $\langle\alpha,\beta\rangle$ in $\mathbf{Con\,A}$ and $\langle\alpha,\beta\rangle$-trace N, then we can assign a type to this trace. The type of the trace N is defined to be the type of the minimal algebra $(\mathbf{A}|_N)/(\alpha|_N)$.

The following theorem will allow us to assign a type to algebras minimal relative to a prime quotient, as well as to $\langle\alpha,\beta\rangle$-minimal sets, and to prime quotients $\langle\alpha,\beta\rangle$.

THEOREM 0.33 *Let* $\mathbf{C}$ *be an* $\langle\alpha,\beta\rangle$*-minimal algebra for some prime quotient* $\langle\alpha,\beta\rangle$ *on* $\mathbf{C}$. *For every pair of* $\langle\alpha,\beta\rangle$*-traces* N_1 *and* N_2, *we have* $N_1 \simeq N_2$ *and* $\mathbf{M}_1 \cong \mathbf{M}_2$ *where* $\mathbf{M}_i = (\mathbf{C}|_{N_i})/(\alpha|_{N_i})$.

PROOF. See the proof of Corollary 5.2 in Hobby, McKenzie. □

Definition 0.34

(1) Let $\mathbf{C}$ be a finite $\langle\alpha,\beta\rangle$-minimal algebra, where $\langle\alpha,\beta\rangle$ is a prime congruence quotient of $\mathbf{C}$. The **type of C relative to** $\langle\alpha,\beta\rangle$ is defined to be the type of any $\langle\alpha,\beta\rangle$-trace in $\mathbf{C}$.

(2) Let $\langle\alpha,\beta\rangle$ be a prime congruence quotient of a finite algebra $\mathbf{A}$. The **type** of $\langle\alpha,\beta\rangle$, written $\mathrm{typ}(\alpha,\beta)$, is the type of any algebra $\mathbf{A}|_U$ relative to $\langle\alpha|_U,\beta|_U\rangle$, where $U \in M_{\mathbf{A}}(\alpha,\beta)$.

(3) Let $\langle\delta,\gamma\rangle$ be any congruence quotient of a finite algebra $\mathbf{A}$.

$$\mathrm{typ}\{\delta,\gamma\} = \{\mathrm{typ}(\alpha,\beta) : \delta \le \alpha \prec \beta \le \gamma\}.$$

(4) For a finite algebra $\mathbf{A}$ we define $\mathrm{typ}\{\mathbf{A}\}$ to be $\mathrm{typ}\{0_A, 1_A\}$.

(5) For a class $\mathcal{K}$ of algebras we define $\mathrm{typ}\{\mathcal{K}\}$ to be the set

$$\bigcup\{\mathrm{typ}\{\mathbf{A}\} : \mathbf{A} \in \mathcal{K} \text{ and } \mathbf{A} \text{ finite}\}.$$

The type of a prime quotient $\langle\alpha,\beta\rangle$ on a finite algebra $\mathbf{A}$ is well defined, since any two $\langle\alpha,\beta\rangle$-minimal sets are polynomially isomorphic in $\mathbf{A}$ and so have the same type.

LEMMA 0.35 *Let* $\delta \le \gamma \le \alpha \prec \beta \le \lambda$ *be congruences on a finite algebra* $\mathbf{A}$. *Then* $\mathrm{typ}(\alpha,\beta) = \mathrm{typ}(\alpha/\delta,\beta/\delta)$ *(computed in* $\mathbf{A}/\delta$*) and* $\mathrm{typ}\{\gamma,\lambda\} = \mathrm{typ}\{\gamma/\delta,\lambda/\delta\}$.

PROOF. See Corollary 5.3 in Hobby, McKenzie. □

The set typ$\{\mathbf{A}\}$ of type labels of a finite algebra $\mathbf{A}$ is a subset of the set $\{\mathbf{1}, \mathbf{2}, \mathbf{3}, \mathbf{4}, \mathbf{5}\}$. This set, and the way the labels are distributed on Con $\mathbf{A}$, determines quite a bit about the structure of the algebra. We say that a congruence quotient $\langle\alpha, \beta\rangle$ is **Abelian** (respectively, **strongly Abelian**, **solvable**, or **strongly solvable**) iff β has the respective property over α. (See §0.4 for the definitions.) Observe that the definitions entail that a prime quotient is solvable iff it is Abelian, and strongly solvable iff it is strongly Abelian. Also, Theorems 0.11 and 0.14 imply that any subquotient of a solvable (strongly solvable) quotient is solvable (strongly solvable).

THEOREM 0.36 *Let $\mathbf{A}$ be a finite algebra, and let $\langle\alpha, \beta\rangle$ be a prime congruence quotient of $\mathbf{A}$, and let $\langle\delta, \gamma\rangle$ be any congruence quotient of $\mathbf{A}$.*

(i) typ$(\alpha, \beta) \in \{\mathbf{1}, \mathbf{2}\}$ *if and only if $\langle\alpha, \beta\rangle$ is Abelian.*

(ii) typ$(\alpha, \beta) = \mathbf{1}$ *if and only if $\langle\alpha, \beta\rangle$ is strongly Abelian.*

(iii) *$\langle\delta, \gamma\rangle$ is solvable (strongly solvable) if and only if* typ$\{\delta, \gamma\} \subseteq \{\mathbf{1}, \mathbf{2}\}$ *(respectively,* typ$\{\delta, \gamma\} = \{\mathbf{1}\}$*).*

PROOF. See the proof of Theorem 5.7 in Hobby, McKenzie. □

LEMMA 0.37 *Let $\langle\alpha_i, \beta_i\rangle$, $i = 1, 2$, be prime congruence quotients of a finite algebra $\mathbf{A}$ that are projective, i.e., $\alpha_2 \wedge \beta_1 = \alpha_1$ and $\alpha_2 \vee \beta_1 = \beta_2$. Then $M_{\mathbf{A}}(\alpha_1, \beta_1) = M_{\mathbf{A}}(\alpha_2, \beta_2)$ and* typ$(\alpha_1, \beta_1) =$ typ(α_2, β_2).

PROOF. See Lemma 6.2 in Hobby, McKenzie. □

LEMMA 0.38 *Let $\langle\alpha, \beta\rangle$ be a prime congruence quotient of a finite algebra $\mathbf{A}$ and let $U \in M_{\mathbf{A}}(\alpha, \beta)$ and B be the $\langle\alpha, \beta\rangle$-body of U. If* typ$(\alpha, \beta) = \mathbf{2}$ *then $\mathbf{A}|_B$ is a Maltsev algebra and $\mathbf{A}|_M$ is Maltsev for every $\langle\alpha, \beta\rangle$-trace $M \subseteq U$. If* typ$(\alpha, \beta) = \mathbf{2}$ *and $\mathbf{A}$ is Abelian then $U = B$.*

PROOF. See Lemmas 4.20 and 4.27 in Hobby, McKenzie. □

Recall that we term a class $\mathcal{K}$ of algebras *unstructured* iff the class of graphs is interpretable into $\mathcal{K}$, and call $\mathcal{K}$ *ω-unstructured* iff the class of finite graphs is interpretable into the class of finite algebras of $\mathcal{K}$. The following theorem summarizes the results of Hobby, McKenzie Chapter 11. Because of this theorem, we shall be concerned in this book only with algebras whose type labels are among $\mathbf{1}, \mathbf{2}$ and $\mathbf{3}$.

THEOREM 0.39 *Let the variety $\mathcal{V}$ be locally finite.*

(i) *If* typ$\{\mathcal{V}\} \not\subseteq \{\mathbf{1}, \mathbf{2}, \mathbf{3}\}$ *then* $\mathcal{V}$ *is unstructured and* ω*-unstructured.*

(ii) *Suppose that* $\mathcal{V}$ *is structured, and let* $\mathbf{A}$ *be a finite algebra in* $\mathcal{V}$ *with a prime quotient* $\langle\alpha, \beta\rangle$ *in* **Con A** *and* U *be an* $\langle\alpha, \beta\rangle$*-minimal set. Then* typ$(\alpha, \beta) = \mathbf{2}$ *implies that* U *has empty tail and* $\mathbf{A}|_U$ *is Maltsev; while* typ$(\alpha, \beta) = \mathbf{3}$ *implies that* U *is an* $\langle\alpha, \beta\rangle$*-trace and* $\mathbf{A}|_U$ *is polynomially equivalent with a two-element Boolean algebra.*

We shall need the following fact about prime quotients of type **1**.

LEMMA 0.40 *Let* $\mathbf{A}$ *be a finite algebra and let* $\alpha \prec \beta$ *in* **Con A** *with* typ$(\alpha, \beta) = \mathbf{1}$. *For every* $f \in \mathrm{Pol}_n\mathbf{A}$ *(for any* n*), and for every set* $T = T_1 \times \cdots \times T_n$, *where* $T_1, \ldots, T_n$ *are* β*-equivalence classes, if* $f(T)$ *is contained in a single* $\langle\alpha, \beta\rangle$*-trace, then* $f|_T$ *depends, modulo* α*, on at most one variable.*

PROOF. See Exercise 5.11(2) or the proof of Theorem 5.6 in Hobby, McKenzie. □

In view of Theorems 0.11 and 0.14, on every finite algebra $\mathbf{A}$ there are congruences α, β, δ and γ such that

(1) α is the largest solvable congruence on $\mathbf{A}$, called the **solvable radical** of $\mathbf{A}$.

(2) β is the smallest congruence on $\mathbf{A}$ such that $\langle\beta, 1_A\rangle$ is solvable, called the **co-solvable radical** of $\mathbf{A}$.

(3) δ is the largest strongly solvable congruence on $\mathbf{A}$, called the **strongly solvable radical** of $\mathbf{A}$.

(4) γ is the smallest congruence such that $\langle\gamma, 1_A\rangle$ is strongly solvable, called the **co-strongly solvable radical** of $\mathbf{A}$.

Here are the last facts about solvability that we shall need.

THEOREM 0.41

(i) *Let* $\mathbf{A}$ *be a finite algebra with* typ$\{\mathbf{A}\} = \{\mathbf{2}\}$. *Then* **Con A** *is a modular lattice.*

(ii) *Let* $\mathcal{V}$ *be a locally finite variety such that* typ$\{\mathcal{V}\} = \{\mathbf{2}\}$. *Then* $\mathcal{V}$ *is Maltsev.*

(iii) *Let* $\mathcal{V}$ *be a locally finite, Abelian variety such that* typ$\{\mathcal{V}\} = \{\mathbf{2}\}$. *Then* $\mathcal{V}$ *is affine.*

PROOF. See Corollary 6.8 and Theorem 7.11 in Hobby, McKenzie for (i) and (ii). (iii) follows from (ii) using Theorem 0.19. □

In this brief presentation we have barely scratched the surface of tame congruence theory. In what follows we do assume a certain amount of fluency in tame congruence theory, and so recommend to the reader to work through some of the proofs in Hobby, McKenzie in order to become familiarized with the theory.

0.7 Definable relations in subdirect powers

A **subdirect power** of an algebra $\mathbf{A}$ is an algebra $\mathbf{D} \leq \mathbf{A}^X$ (for some set X) such that the image of $\mathbf{D}$ under each coordinate projection is $\mathbf{A}$. In all our work, we shall be looking for ways to interpret a class of graphs, or Boolean pairs, into the class $\boldsymbol{P}_s(\mathbf{A})$ of subdirect powers of a finite algebra $\mathbf{A}$. (See §0.2.) To accomplish this, we must study the definable relations in subdirect powers of $\mathbf{A}$.

Definition 0.42 Let $\mathbf{D}$ be any algebra or structure. A subset $r \subseteq D^n$ is called **definable in D** iff there is a formula $\phi(\bar{x}, \bar{y})$ in the language of $\mathbf{D}$ with free variables $\bar{x} = \langle x_0, \ldots, x_{n-1}\rangle$ and $\bar{y} = \langle y_0, \ldots, y_{m-1}\rangle$ (for some m) and there is a sequence $\bar{d} \in D^m$ such that

$$r = \phi^{\langle \mathbf{D}, \bar{d}\rangle} = \{\bar{c} \in D^n : \mathbf{D} \models \phi(\bar{c}, \bar{d})\}.$$

When this holds, we say that r is defined in $\mathbf{D}$ by the formula $\phi(\bar{x}, \bar{y})$ with the parameters $\bar{d}$.

Good candidates for definable relations in an algebra $\mathbf{D} \leq \mathbf{A}^X$ are the factorable relations corresponding to relations definable in $\mathbf{A}$.

Definition 0.43 Let $\mathbf{A}$ be an algebra, X be a set, $\phi(x_0, \ldots, x_{n-1})$ be a formula in the language of $\mathbf{A}$, $r \subseteq A^n$ be an n-ary relation on A, and $f_0, \ldots, f_{n-1}$ be elements of A^X. We put

$$[\![r(f_0, \ldots, f_{n-1})]\!] = \{x \in X \;:\; \langle f_0(x), \ldots, f_{n-1}(x)\rangle \in r\},$$

$$[\![\phi(f_0, \ldots, f_{n-1})]\!] = \{x \in X \;:\; \mathbf{A} \models \phi(f_0(x), \ldots, f_{n-1}(x))\}.$$

Suppose that $D \subseteq A^X$. Then we put

$$D(r) = \{\langle f_0, \ldots, f_{n-1}\rangle \in D^n \;:\; [\![r(f_0, \ldots, f_{n-1})]\!] = X\},$$

$$D(\phi) = \{\langle f_0, \ldots, f_{n-1}\rangle \in D^n \;:\; [\![\phi(f_0, \ldots, f_{n-1})]\!] = X\}.$$

A relation $s \subseteq D^n$ is called **factorable** iff there are relations $s^{(x)}$ on A for all $x \in X$ (the **factors** of s) such that $\langle f_0, \ldots, f_{n-1} \rangle \in s$ is equivalent to $\{f_0, \ldots, f_{n-1}\} \subseteq D$ and $\langle f_0(x), \ldots, f_{n-1}(x) \rangle \in s^{(x)}$ for all $x \in X$. (Note that a relation of the kind $s = D(r)$ ($r \subseteq A^n$) is factorable. Also, if $r \subseteq A^{n+m}$ and $s = D(r)$ and $\bar{g} \in D^m$ then $s^{\langle D, \bar{g} \rangle} = \{\bar{f} \in D^n : \langle \bar{f}, \bar{g} \rangle \in s\}$ is factorable, with factors $r^{\langle A, \bar{g}(x) \rangle}$ ($x \in X$).)

Given $f, g \in A^X$, the **equalizer** of f and g is the set $[\![f = g]\!]$, i.e., the set of all $x \in X$ such that $f(x) = g(x)$. The **co-equalizer** of f and g is the set $[\![f \neq g]\!]$ of all $x \in X$ such that $f(x) \neq g(x)$. We say that f and g are **almost equal**, and we write $f \overset{\text{ae}}{=} g$, iff $[\![f \neq g]\!]$ is a finite set. A set $P \subseteq A^X$ is called **ae-closed** iff for all $f \in P$ and $g \in A^X$, if $f \overset{\text{ae}}{=} g$ then $g \in P$. The **ae-closure** of P is the set of all $g \in A^X$ such that $g \overset{\text{ae}}{=} f$ for some $f \in P$. We remark that the ae-closure of any subuniverse of $\mathbf{A}^X$ is a subuniverse.

The constant function in A^X corresponding to an element $e \in A$ will be denoted $\hat{e}$. Thus $\hat{e}(x) = e$ for all $x \in X$. An algebra $\mathbf{D} \leq \mathbf{A}^X$ is called a **diagonal subalgebra** of $\mathbf{A}^X$ iff $\hat{e} \in D$ for all $e \in A$. Diagonal subalgebras are subdirect powers of $\mathbf{A}$, and may be called **diagonal subdirect powers** of $\mathbf{A}$.

For an n-ary operation f on A, we denote by $f^{(X)}$ the n-ary operation on A^X such that $f^{(X)}(\bar{u})(x) = f(\bar{u}(x))$, and we say that $f^{(X)}$ **acts coordinate-wise like** f.

If $r \subseteq A^n$ is the relation defined by a formula $\phi(\bar{x}, \bar{d})$, i.e., if $r = \phi^{\langle \mathbf{A}, \bar{d} \rangle}$, and if $\mathbf{D} \leq \mathbf{A}^X$ is a diagonal subalgebra, then it will often (but not always) be the case that the factorable relation $D(r)$ (defined above) is defined in $\mathbf{D}$ by the formula $\phi(\bar{x}, \bar{d}')$ where $\bar{d} = \langle d_0, \ldots, d_{m-1} \rangle$ and $\bar{d}' = \langle \hat{d}_0, \ldots, \hat{d}_{m-1} \rangle$. This will always be the case if r is the graph of an $n-1$-ary polynomial operation $t^{\mathbf{A}}(\bar{x}, \bar{a})$ of $\mathbf{A}$ (t a term), so that r is defined by the formula $x_{n-1} \approx t(x_0, \ldots, x_{n-2}, \bar{a})$. $D(r)$ is in this case the graph of the polynomial operation $t^{\mathbf{D}}(\bar{x}, \bar{a}')$ of $\mathbf{D}$.

Note that if $f = t^{\mathbf{A}}$ is the term operation corresponding to a term t, and if D is any subuniverse of $\mathbf{A}^X$, we have $f^{(X)}|_D = t^{\mathbf{D}}$. If $f(\bar{x}) = t^{\mathbf{A}}(\bar{x}, \bar{a})$ is a polynomial operation of $\mathbf{A}$ (where $\bar{a} \in A^m$ for some m), and if $\mathbf{D}$ is a diagonal subalgebra of $\mathbf{A}$, then $f^{(X)}|_D = t^{\mathbf{D}}(\bar{x}, \bar{a}')$ is a polynomial operation of $\mathbf{D}$ where $\bar{a}' = \langle \hat{a}_0, \ldots, \hat{a}_{m-1} \rangle$. We shall often use these facts, and the following elementary results concerning polynomial operations in subalgebras of direct products.

LEMMA 0.44

(i) *Let* $\mathbf{D} \leq \prod \langle \mathbf{A}_x : x \in X \rangle$. *Let* $p(\bar{z}) = t^{\mathbf{D}}(\bar{z}, \bar{g})$ *be an* n*-ary polynomial operation of* $\mathbf{D}$, *where* t *is a term and* $\bar{g} \in D^m$ *for some* m. *Then*

p is a factorable operation with factors the polynomial operations $p^{(x)}(\bar{z}) = t^{\mathbf{A}_x}(\bar{z}, \bar{g}(x))$ $(x \in X)$ of $\mathbf{A}_x$; i.e., $p(\bar{f})(x) = p^{(x)}(\bar{f}(x))$ for all $\bar{f} \in D^n$ and all $x \in X$.

(ii) *An algebra $\mathbf{D} \leq \mathbf{A}^X$ is a diagonal subalgebra of $\mathbf{A}^X$ iff $\mathbf{D}$ is closed under $p^{(X)}$ for every polynomial operation p of $\mathbf{A}$.*

Definition 0.45 Let X be a set and $\mathbf{B} \subseteq \mathcal{P}(X)$ be a Boolean subalgebra of the algebra of all subsets of X. Let $\mathbf{A}$ be any algebra. The **Boolean power** of $\mathbf{A}$ by $\mathbf{B}$, denoted $\mathbf{A}[\mathbf{B}]$, is the subalgebra of $\mathbf{A}^X$ whose elements are all the functions $f \in A^X$ satisfying: $f(X)$ is finite and for all $a \in A$, $f^{-1}(a) \in B$.

$\mathbf{A}[\mathbf{B}]$ is isomorphic to the algebra of continuous functions from the Stone space of $\mathbf{B}$ to A (where A is given the discrete topology). Thus within isomorphism, $\mathbf{A}[\mathbf{B}]$ depends only upon $\mathbf{A}$ and the isomorphism class of $\mathbf{B}$. This construction (the Boolean power) is not immediately useful for our purposes; in fact, if $\mathbf{A}$ is finite then the first order theory of the class $\{\mathbf{A}[\mathbf{B}] : \mathbf{B}$ a Boolean algebra$\}$ is decidable. However, there are many ways to modify the construction; and classes of modified Boolean powers of a finite algebra often prove to be undecidable (and unstructured). A varied assortment of these modified Boolean powers will be used in the book. In dealing with them, we employ all of the notation introduced in this section. Notice that $\mathbf{A}[\mathbf{B}]$ is a diagonal subdirect power of $\mathbf{A}$; and it is ae-closed in $\mathbf{A}^X$ if $\mathbf{B}$ includes every finite subset of X as a member.

Definition 0.46 $\mathcal{BP}_1$ is the class of structures $\langle B_1, B_0, \leq\rangle$ such that for some nonvoid set X, B_0 and B_1 are Boolean subuniverses of the algebra $\mathcal{P}(X)$ of all subsets of X, $B_0 \subseteq B_1$, every finite subset of X belongs to B_0, and $\leq$ is the Boolean order on B_1. Such structures are called **Boolean pairs** (of the first kind).

THEOREM 0.47 *$\mathcal{BP}_1$ is an unstructured class.*

PROOF. See Theorems 6.1 and 6.2 in Burris, McKenzie [1981] where it is shown that the class of finite graphs is interpretable into $\mathcal{BP}_1$. Using the same construction and formulas from their paper, and introducing minor changes into the arguments, it can be shown that the class of all graphs is interpretable into $\mathcal{BP}_1$. □

The next theorem exemplifies the use of modified Boolean powers to interpret $\mathcal{BP}_1$ into the class of subdirect powers of an algebra. The theorem is modelled on Theorem 7.12 in Burris and McKenzie. The result will be used in Chapter 3.

THEOREM 0.48 *Let* $\mathbf{F}$ *be a finite algebra for which there is a formula* $P_{\subseteq}(x, y, z, w)$ *satisfying: whenever* $\mathbf{D}$ *is an ae-closed subdirect power of* $\mathbf{F}$ *and* $\{f, g, h, k\} \subseteq D$, *then*

$$\mathbf{D} \models P_{\subseteq}(f, g, h, k) \text{ iff } [\![f = g]\!] \subseteq [\![h = k]\!] .$$

Suppose that $\mathbf{F}$ *has a subalgebra which possesses three or more congruences. Then the class* $\boldsymbol{P}_s(\mathbf{F})$ *is unstructured.*

PROOF. Let $\mathbf{S}$ be a subalgebra of $\mathbf{F}$ and γ be a congruence on $\mathbf{S}$ with $0_S < \gamma < 1_S$. Choose a γ-equivalence class in $\mathbf{S}$ with more than one element and enumerate it as $c_0, \ldots, c_{k-1}$, so that $c_0/\gamma = \{c_0, \ldots, c_{k-1}\}$. Choose an element $a \in S$ such that $a \notin c_0/\gamma$.

We shall prove this lemma by interpreting the class $\mathcal{BP}_1$ into $\boldsymbol{P}_s(\mathbf{F})$. Let $\mathbf{B} = \langle B_1, B_0, \leq \rangle$ be a member of $\mathcal{BP}_1$ with $B_1 \subseteq \mathcal{P}(X)$. The algebra $\mathbf{D} \in \boldsymbol{P}_s(\mathbf{F})$ in which $\mathbf{B}$ will be interpreted is defined in the following way.

Let $\mathbf{D}' = \mathbf{S}[B_0, B_1, \gamma]$ be the subalgebra of $\mathbf{S}^X$ consisting of all the functions f such that $f^{-1}(s) \in B_1$ and $f^{-1}(s/\gamma) \in B_0$ for all $s \in S$. Then let $\mathbf{D}$ be the ae-closure of $\mathbf{D}'$ in $\mathbf{F}^X$.

Notice that $\mathbf{D}'$ is an ae-closed diagonal subalgebra of $\mathbf{S}^X$, since B_0 contains all the finite subsets of X. $\mathbf{D}$ is an ae-closed subalgebra of $\mathbf{F}^X$; but it is not diagonal unless X is finite or $\mathbf{S} = \mathbf{F}$.

In order to interpret $\mathbf{B}$, we need a scheme $\bar{\psi} = \langle \mathrm{Un}, \mathrm{B}_0, \mathrm{Eq}, \mathrm{P}_{\leq} \rangle$ consisting of four formulas. We shall use the constant functions $\hat{c}_i$ as parameters. The sequence of variables used as parameters in our formulas will be $\bar{y} = \langle y_0, \ldots, y_{k-1} \rangle$, and they will be interpreted by the sequence $\bar{c}' = \langle \hat{c}_0, \ldots, \hat{c}_{k-1} \rangle$. Our interpretation will be motivated by the observation that the mapping φ defined by $\varphi(f, g) = [\![f = g]\!]$ maps D^2 onto B_1. (Since $\langle c_0, c_1 \rangle \in \gamma - 0_S$, for every set $Y \in B_1$ the function f such that $f(Y) = \{c_1\}$ and $f(X - Y) = \{c_0\}$ belongs to D' and $\varphi(\hat{c}_1, f) = Y$.) Thus the elements of B_1 will be interpreted by arbitrary pairs of elements of D, modulo an equivalence relation. Three of the formulas we need are easily produced.

$$\begin{aligned}
\mathrm{Un}(x_0, x_1, \bar{y}) &\overset{\text{def}}{\longleftrightarrow} (x_0 \approx x_0) \\
\mathrm{P}_{\leq}(x_0^1, x_1^1, x_0^2, x_1^2, \bar{y}) &\overset{\text{def}}{\longleftrightarrow} \mathrm{P}_{\subseteq}(x_0^1, x_1^1, x_0^2, x_1^2) \\
\mathrm{Eq}(x_0^1, x_1^1, x_0^2, x_1^2, \bar{y}) &\overset{\text{def}}{\longleftrightarrow} \mathrm{P}_{\leq}(x_0^1, x_1^1, x_0^2, x_1^2) \wedge \mathrm{P}_{\leq}(x_0^2, x_1^2, x_0^1, x_1^1) .
\end{aligned}$$

Clearly, we have, for any $f, g, h, k \in D$:

$$\varphi(f, g) \leq \varphi(h, k) \text{ iff } \mathbf{D} \models \mathrm{P}_{\leq}(f, g, h, k) \text{ and}$$

$$\varphi(f,g) = \varphi(h,k) \text{ iff } \mathbf{D} \models \mathrm{Eq}(f,g,h,k).$$

All that remains now is to find a formula defining in $\mathbf{D}$ the binary relation $\varphi^{-1}(B_0)$. This is where the parameters come in. We can take

$$\mathrm{B}_0(x_0,x_1,y_0,\ldots,y_{k-1}) \overset{\text{def}}{\longleftrightarrow} (\exists z)\Big[\Big\{\bigwedge_{i<k} \mathrm{P}_{\le}(z,y_i,x_0,x_1)\Big\} \wedge$$

$$\Big\{(\forall u,v)[(\bigwedge_{i<k} \mathrm{P}_{\le}(z,y_i,u,v)) \to \mathrm{P}_{\le}(x_0,x_1,u,v)]\Big\}\Big].$$

It is easily verified that $\mathbf{D} \models \mathrm{B}_0(f,g,\bar{c}')$ iff there exists $h \in D$ such that $[\![f = g]\!] = h^{-1}(c_0/\gamma)$. Now for any $h \in D$ we have $h \overset{\text{ae}}{\equiv} h'$ for some $h' \in D'$, and for such an h' the sets $h^{-1}(c_0/\gamma)$ and $h'^{-1}(c_0/\gamma)$ differ by a finite set. Thus $h^{-1}(c_0/\gamma) \in B_0$ for all $h \in D$; and we have that $\mathbf{D} \models \mathrm{B}_0(f,g,\bar{c}')$ implies $\varphi(f,g) \in B_0$. Conversely, if $f,g \in D$ and $[\![f = g]\!] = Y \in B_0$, then the function h such that $h(Y) = \{c_0\}$ and $h(X - Y) = \{a\}$ belongs to D' and $h^{-1}(c_0/\gamma) = Y$; thus $\mathbf{D} \models \mathrm{B}_0(f,g,\bar{c}')$.

This concludes both our construction of the scheme $\bar{\psi}$ for interpreting $\mathcal{BP}_1$ into $\boldsymbol{P}_s(\mathbf{F})$, and our construction, for each $\mathbf{B} \in \mathcal{BP}_1$, of an algebra $\mathbf{D} \in \boldsymbol{P}_s(\mathbf{F})$ and a k-tuple $\bar{d}$ of elements of $\mathbf{D}$ such that $\mathbf{B} \cong \mathbf{D}(\bar{\psi},\bar{d})$. We have shown that $\mathcal{BP}_1$ is interpretable into $\boldsymbol{P}_s(\mathbf{F})$. □

We conclude this Chapter by proving a result that will be needed in Chapter 13. The proof illustrates one of the ideas used by Hobby and McKenzie in proving Theorem 0.39. Recall from the second paragraph of §0.6 that we use the notation $\mathbf{A}|_U$, where U is a nonvoid subset of an algebra $\mathbf{A}$, to denote the non-indexed algebra $\langle U, (\mathrm{Pol}\ \mathbf{A})|_U\rangle$. For Theorem 0.50, we need an indexed algebra equivalent to $\mathbf{A}|_U$.

Definition 0.49 $\mathbf{A}\mathrm{I}_U$ is the indexed algebra $\langle U, f(f \in (\mathrm{Pol}\ \mathbf{A})|_U)\rangle$ with language $(\Phi, \emptyset, \rho)$ where $\rho(f) = n$ if $f \in (\mathrm{Pol}_n \mathbf{A})|_U$.

If $\mathsf{L} = (\Phi, \Sigma, \rho)$ is a language, by a **sublanguage** of L we mean a language $\mathsf{L}' = (\Phi', \Sigma', \rho')$ such that $\Phi' \subseteq \Phi$, $\Sigma' \subseteq \Sigma$ and $\rho' \subseteq \rho$. If L' is a sublanguage of L and $\mathbf{A} \in \mathrm{Mod}(\mathsf{L})$, then by the **reduct of A** to L' we mean the L'-structure $\mathbf{A}|_{\mathsf{L}'} = \langle A, f^{\mathbf{A}}(f \in \Phi'), s^{\mathbf{A}}(s \in \Sigma')\rangle$. If $\mathcal{K} \subseteq \mathrm{Mod}(\mathsf{L})$, then by $\mathcal{K}|_{\mathsf{L}'}$ we mean the class of all reducts $\mathbf{A}|_{\mathsf{L}'}$ where $\mathbf{A} \in \mathcal{K}$. Note that if a class $\mathcal{K}'$ of L'-structures is interpretable into a class $\mathcal{K}$ of L-structures (where L' is a finite language, according to Definition 0.1), then $\mathcal{K}'$ is interpretable into a reduct $\mathcal{K}|_{\mathsf{L}''}$ for some finite sublanguage L'' of L; and conversely, any class interpretable into a reduct of $\mathcal{K}$ is interpretable into $\mathcal{K}$. Finally, recall that where $\mathbf{A}$ is an algebra, $\mathrm{E}(\mathbf{A})$ denotes the set of unary polynomial operations e of $\mathbf{A}$ such that $e = e \circ e$.

THEOREM 0.50 *Let* $\mathbf{A}$ *be an algebra,* $e \in \mathrm{E}(\mathbf{A})$*, and* $U = e(A)$*. Let* $\mathcal{V}_0 = V(\mathbf{A}|_U)$ *and* $\mathcal{V}_1 = V(\mathbf{A})$ *be the varieties generated by* $\mathbf{A}|_U$ *and* $\mathbf{A}$*, respectively, and let* L_0 *be the language of* $\mathcal{V}_0$*. Then for every finite language* $\mathsf{L} \subseteq \mathsf{L}_0$*,* $\mathcal{V}_0|_{\mathsf{L}}$ *is interpretable into* $\mathcal{V}_1$*.*

PROOF. Let the operation symbols of $\mathsf{L} \subseteq \mathsf{L}_0$ be $f_1, \ldots, f_k$; and put $\mathcal{K} = \mathcal{V}_0|_{\mathsf{L}}$. Since $f_1, \ldots, f_k \in (\mathrm{Pol}\mathbf{A})|_U$, we can choose $g_1, \ldots, g_k \in \mathrm{Pol}\,\mathbf{A}$ such that $g_a|_U = f_a$ $(1 \leq a \leq k)$. We can then choose (for some n and variables $\bar{y} = \langle y_1, \ldots, y_n \rangle$) terms $\varepsilon(x, \bar{y})$, $\tau_1(\bar{x}_1, \bar{y}), \ldots, \tau_k(\bar{x}_k, \bar{y})$ and a sequence $\bar{a} \in A^n$ so that

$$\begin{aligned} e(x) &= \varepsilon^{\mathbf{A}}(x, \bar{a}), \\ g_1(\bar{x}_1) &= \tau_1^{\mathbf{A}}(\bar{x}_1, \bar{a}), \\ &\vdots \\ g_k(\bar{x}_k) &= \tau_k^{\mathbf{A}}(\bar{x}_k, \bar{a}). \end{aligned}$$

For our interpretation we use the scheme

$$\bar{\psi} = \langle \mathrm{Un}, \mathrm{Eq}, F_1, \ldots, F_k \rangle$$

where

$$\begin{aligned} \mathrm{Un}(x, \bar{y}) &\overset{\mathrm{def}}{\longleftrightarrow} \varepsilon(x, \bar{y}) \approx x, \\ \mathrm{Eq}(x_1, x_2, \bar{y}) &\overset{\mathrm{def}}{\longleftrightarrow} x_1 \approx x_2, \\ F_a(x_1, \ldots, x_{p+1}, \bar{y}) &\overset{\mathrm{def}}{\longleftrightarrow} \tau_a(x_1, \ldots, x_p, \bar{y}) \approx x_{p+1}, \end{aligned}$$

(if f_a is p-ary).

We claim of course that $\bar{\psi}$ interprets $\mathcal{K}$ into $\mathcal{V}_1$. (See Definition 0.1.) Due to the simple nature of the scheme we have chosen, our claim reduces to this: For every $\mathbf{D}' \in \mathcal{K}$ there exists $\mathbf{D} \in \mathcal{V}_1$ and $\bar{d} \in D^n$ so that where $h(x) = \varepsilon^{\mathbf{D}}(x, \bar{d})$ and $h_a(\bar{x}) = \tau_a^{\mathbf{D}}(\bar{x}, \bar{d})$ $(1 \leq a \leq k)$, we have that the set $Q = \{x \in D : h(x) = x\}$ is nonvoid, Q is closed under each of the operations h_a, and $\mathbf{D}' \cong \langle Q, h_1|_Q, \ldots, h_k|_Q \rangle$.

To prove the claim, let $\mathbf{D}' \in \mathcal{K}$. Thus $\mathbf{D}' = \mathbf{E}|_{\mathsf{L}}$ for some $\mathbf{E} \in \mathcal{V}_0$. By definition of $\mathcal{V}_0$, there is a set X, an algebra $\mathbf{S} \leq (\mathbf{A}|_U)^X$, and an epimorphism $\chi : \mathbf{S} \to \mathbf{E}$. Now $S \subseteq A^X$, and we let $\mathbf{B} \leq \mathbf{A}^X$ be the subalgebra of $\mathbf{A}^X$ generated by $S \cup \Delta$ where Δ is the set of all constant functions $\hat{a}$ for $a \in A$.

In the notation introduced in Definition 0.43 (the case of a 1-ary relation r), we have $B(U) = B \cap U^X$. Let $\bar{a}'$ be the sequence of diagonal elements corresponding to the sequence $\bar{a} \in A^n$. Then we have $e^{(X)}(z) = \varepsilon^{\mathbf{B}}(z, \bar{a}')$

for $z \in B$. Since e is idempotent, it is clear that $e^{(X)}$ is idempotent and that $e^{(X)}(B) = B(e(A)) = B(U)$. Let us write simply e for the polynomial $e^{(X)}|_B$ of $\mathbf{B}$. Now we claim that $e(B)$ $(= B(U)) = S$. Certainly $S \subseteq B \cap U^X$ by definition, and so $S \subseteq e(B)$. On the other hand, let $u \in e(B)$, so that $e(u) = u$. From the definition of B it follows that there is a term σ and a sequence $\bar{b}' = \langle \hat{b}_1, \ldots, \hat{b}_l \rangle \in \Delta^l$ and a sequence $\bar{s} \in S^m$ such that $u = \sigma^{\mathbf{B}}(\bar{s}, \bar{b}')$. Now $p(\bar{z}) = e\sigma^{\mathbf{A}}(\bar{z}, \bar{b})$ is a polynomial of $\mathbf{A}$ (where $\bar{b} = \langle b_1, \ldots, b_l \rangle$), and U is closed under p. Thus $f = p|_U$ is one of the basic operations of $\mathbf{A}|_U$. Hence $f^{\mathbf{S}} = p^{(X)}|_S$ is one of the basic operations of $\mathbf{S}$. We have

$$u = e(u) = e\sigma^{\mathbf{B}}(\bar{s}, \bar{b}') = f^{\mathbf{S}}(\bar{s})$$

showing that $u \in S$ as desired. Thus $e(B) = S$ as claimed.

We next claim that $\mathbf{B}|_S = \mathbf{S}|_S$—i.e., that the polynomials of $\mathbf{S}$ are the same as the restrictions to S of the polynomials of $\mathbf{B}$ under which S is closed. The (straightforward) proof of this claim we leave to the reader.

Now let $\gamma = \ker \chi$. Then γ is a congruence on $\mathbf{S}$, and thus also on $\mathbf{B}|_S$. We next define

$$\Gamma = \left\{ \langle u, v \rangle \in B^2 : (\forall f \in \operatorname{Pol}_1 \mathbf{B})(\langle ef(u), ef(v) \rangle \in \gamma) \right\}.$$

It is easy to see that Γ is a congruence on $\mathbf{B}$ and that $\Gamma|_S = \gamma$ (i.e., γ is the restriction to S of Γ). (For proving the latter assertion, use the fact that $ef|_S \in \operatorname{Pol}_1 \mathbf{B}|_S$ for all $f \in \operatorname{Pol}_1 \mathbf{B}$.)

We now define $\mathbf{D} = \mathbf{B}/\Gamma$ and take $\bar{d} = \langle d_1, \ldots, d_n \rangle \in D^n$ where $d_i = \hat{a}_i/\Gamma$. Since χ is a surjective homomorphism $\mathbf{S} \to \mathbf{E}$ and $\Gamma|_S = \ker \chi$, we have a bijection ϕ from the set $S/\Gamma \subseteq D$ onto E satisfying $\phi(s/\Gamma) = \chi(s)$. We claim now that $S/\Gamma = \operatorname{Un}^{\langle \mathbf{D}, \bar{d} \rangle}$. The truth of this claim becomes clear when we observe that for $w = u/\Gamma$ ($u \in B$), we have

$$\begin{aligned} \mathbf{D} \models \operatorname{Un}(w, \bar{d}) &\longleftrightarrow \varepsilon^{\mathbf{B}}(u, \bar{a}') \equiv_\Gamma u \\ &\longleftrightarrow u \in e(u)/\Gamma \\ &\longleftrightarrow u \in \bigcup (e(B)/\Gamma) \\ &\longleftrightarrow w \in S/\Gamma. \end{aligned}$$

The remainder of the verification that $\phi : \mathbf{D}(\bar{\psi}, \bar{d}) \cong \mathbf{D}'$ is left to the reader. The major step is just to show that $f_i^{\mathbf{S}}(\bar{z}) = \tau_i^{\mathbf{B}}(\bar{z}, \bar{a}')$, for $1 \le i \le k$ and $\bar{z} \in S^p$ (where f_i is p-ary). □

Chapter 1

Preview: The three subvarieties

The aim of this work is to prove that every locally finite variety that fails to be hereditarily undecidable decomposes as the product of three subvarieties which are, respectively, strongly Abelian, affine, and a discriminator variety.

A variety that is not hereditarily undecidable must be both structured and ω-structured (see Definition 0.4). Our decomposition theorem will be proved for locally finite varieties that are structured; but some of the intermediate results will be seen to hold also for varieties that are ω-structured. We define at the outset three subvarieties, $\mathcal{V}_1$, $\mathcal{V}_2$ and $\mathcal{V}_3$, of a locally finite variety $\mathcal{V}$; and we prove that if $\mathcal{V}$ is structured, then every member of $\mathcal{V}$ is a subdirect product of three algebras belonging to these subvarieties.

The chief results in Part I will be proved to hold for every structured locally finite variety $\mathcal{V}$. They are: (Theorem 5.4) the subvariety $\mathcal{V}_1 \vee \mathcal{V}_2$ is Abelian; and (Theorem 4.1) $\mathcal{V}_3$ is a discriminator variety.

In Part II, we deal with an arbitrary structured locally finite, Abelian variety $\mathcal{V}$. It is trivial to see that $\mathcal{V} = \mathcal{V}_1 \vee \mathcal{V}_2$; in fact, every algebra in $\mathcal{V}$ is a subdirect product of an algebra in $\mathcal{V}_1$ and an algebra in $\mathcal{V}_2$. We shall prove that $\mathcal{V}_1$ is strongly Abelian and that $\mathcal{V}_2$ is affine (Theorem 9.6). We shall then present an effective characterization of the structured locally finite, strongly Abelian varieties (Corollary 12.17, Corollary 12.18, and Theorem 12.19).

In Part III, we deal again with an arbitrary structured locally finite variety $\mathcal{V}$. Making use of the results of Parts I and II, we will prove that $\mathcal{V}$ decomposes as the product of its subvarieties $\mathcal{V}_1$, $\mathcal{V}_2$ and $\mathcal{V}_3$ (Theorem 13.10). Some interesting corollaries that follow from our results, together with a list of open problems, can be found in Chapter 14.

The following definition will be used in all three parts of the argument.

Definition 1.1 Let $\mathcal{V}$ be any locally finite variety. For $1 \leq i \leq 3$, let $\mathcal{S}_i$ be the class of all finite, subdirectly irreducible algebras $\mathbf{A} \in \mathcal{V}$ such that the type of $\langle 0_A, \beta \rangle$ is $\mathbf{i}$, where β is the monolith of $\mathbf{A}$. We define $\mathcal{V}_i$ to be the variety generated by $\mathcal{S}_i$.

Here are three easy consequences of the definition.

THEOREM 1.2 *Let $\mathcal{V}$ be a locally finite variety. If $\mathcal{V}$ is Abelian, then every finite subdirectly irreducible algebra in $\mathcal{V}$ belongs to $\mathcal{V}_1 \cup \mathcal{V}_2$, and $\mathcal{V}_3$ is the trivial variety of one-element algebras. If $\mathcal{V}$ is structured, every finite subdirectly irreducible algebra in $\mathcal{V}$ belongs to $\mathcal{V}_1 \cup \mathcal{V}_2 \cup \mathcal{V}_3$.*

PROOF. If $\mathbf{A}$ is a finite, subdirectly irreducible, Abelian algebra in $\mathcal{V}$, then $\text{typ}\{\mathbf{A}\} \subseteq \{\mathbf{1}, \mathbf{2}\}$; hence $\mathbf{A} \in \mathcal{S}_1 \cup \mathcal{S}_2$. If $\mathcal{V}$ is structured, then it follows from Theorem 0.39 that every finite, subdirectly irreducible algebra in $\mathcal{V}$ belongs to $\mathcal{S}_1 \cup \mathcal{S}_2 \cup \mathcal{S}_3$. □

THEOREM 1.3 *Let $\mathcal{V}$ be a locally finite variety. Every locally solvable algebra in $\mathcal{V}$ belongs to $\mathcal{V}_1 \vee \mathcal{V}_2$.*

PROOF. Let $\mathbf{A} \in \mathcal{V}$ be locally solvable. Since $\mathcal{V}$ is locally finite, every finitely generated subalgebra of $\mathbf{A}$ is finite; and so $\mathbf{A}$ belongs to the variety generated by its finite subalgebras. Hence it suffices to prove that all finite subalgebras of $\mathbf{A}$ belong to $\mathcal{V}_1 \vee \mathcal{V}_2$. Let $\mathbf{B}$ be a finite subalgebra of $\mathbf{A}$. $\mathbf{B}$ is a subdirect product of subdirectly irreducible homomorphic images of $\mathbf{B}$, which are solvable, and hence have monoliths of types $\mathbf{1}$ or $\mathbf{2}$. Thus $\mathbf{B} \in \boldsymbol{SP}_{\text{fin}}(\mathcal{S}_1 \cup \mathcal{S}_2)$, implying that $\mathbf{B} \in \mathcal{V}_1 \vee \mathcal{V}_2$. □

THEOREM 1.4 *Let $\mathcal{V}$ be a locally finite variety and assume that either $\mathcal{V}$ is Abelian, or $\mathcal{V}$ is structured. Then $\mathcal{V} = \mathcal{V}_1 \vee \mathcal{V}_2 \vee \mathcal{V}_3$; in fact, every algebra in $\mathcal{V}$ is a subdirect product of three algebras $\mathbf{C}_1, \mathbf{C}_2$ and $\mathbf{C}_3$ belonging, respectively, to $\mathcal{V}_1$, $\mathcal{V}_2$ and $\mathcal{V}_3$.*

PROOF. It follows easily from Theorem 1.2 that every finite algebra in $\mathcal{V}$ can be embedded into a product $\mathbf{C}_1 \times \mathbf{C}_2 \times \mathbf{C}_3$ with $\mathbf{C}_i \in \mathcal{V}_i$. The class of algebras that can be so embedded is closed under the formation of ultraproducts and of subalgebras. The theorem thus follows from the fact that every locally finite algebra can be embedded into an ultraproduct of its finite subalgebras. □

Part I

Structured varieties

Chapter 2

A property of the center

Throughout this chapter, and indeed throughout Part I, $\mathcal{V}$ denotes a fixed, but arbitrary, structured locally finite variety.

The center of any algebra $\mathbf{A}$ is a congruence, $Z(\mathbf{A})$, which we defined in Definition 0.7. Recall that $\mathbf{A}$ is called *centerless* iff $Z(\mathbf{A}) = 0_A$. The next theorem has a very long proof, which is the entire content of this chapter.

THEOREM 2.1 *If* $\mathbf{A}$ *is any finite algebra in* $\mathcal{V}$*, then* $\mathbf{A}/Z(\mathbf{A})$ *is a centerless algebra.*

Combined with the results of the next chapter, Theorem 2.1 will imply that the center of any algebra in $\mathcal{V}$ is the largest locally solvable congruence of the algebra, and thus that every locally solvable algebra in $\mathcal{V}$ is Abelian.

Most of the intermediate results proved in Part I, like Theorem 2.1, assert that $\mathcal{V}$ has no finite algebra possessing a given property. In each case, the theorem is proved by showing that if $\mathbf{A}$ is any finite algebra possessing the property in question, then either the class of graphs or some class of structures in which the class of graphs can be interpreted, is interpretable in the class of subdirect powers of $\mathbf{A}$, and hence in $\mathcal{V}$. Since we are assuming that $\mathcal{V}$ is structured, it follows that $\mathbf{A} \notin \mathcal{V}$. (See Definition 0.4.) We begin the proof of Theorem 2.1 with a definition.

Definition 2.2 Let $\mathbf{A}$ be an algebra, t be an $n+1$ -ary term in the language of $\mathbf{A}$, and $\bar{c}, \bar{d} \in A^n$. We denote by $E^{\mathbf{A}}(t, \bar{c}, \bar{d})$ the set of all $a \in A$ such that $t^{\mathbf{A}}(a, \bar{c}) = t^{\mathbf{A}}(a, \bar{d})$

Recall that we defined in Definition 0.43 the concepts of a diagonal subalgebra of $\mathbf{A}^X$ and an ae-closed subalgebra of $\mathbf{A}^X$, and we defined a relation $D(\rho)$ on D corresponding to any relation ρ on A, where $\mathbf{D}$ is any subalgebra of $\mathbf{A}^X$. In this chapter, we shall often write $Z_1(\mathbf{A})$ in place of

$Z(\mathbf{A})$ for the center of $\mathbf{A}$, and then we write $Z_2(\mathbf{A})$ for the congruence of $\mathbf{A}$ containing $Z_1(\mathbf{A})$ such that $Z_2(\mathbf{A})/Z_1(\mathbf{A})$ is the center of $\mathbf{A}/Z_1(\mathbf{A})$. Note that $\mathbf{A}/Z(\mathbf{A})$ is centerless iff $Z_1(\mathbf{A}) = Z_2(\mathbf{A})$.

LEMMA 2.3 *Let $\mathbf{A}$ be a finite algebra. There exist two first order formulas which define $D(Z_1(\mathbf{A}))$ and $D(Z_2(\mathbf{A}))$ whenever $\mathbf{D}$ is an ae-closed subdirect power of $\mathbf{A}$. Moreover, for such an algebra $\mathbf{D}$, $Z_i(\mathbf{D}) = D(Z_i(\mathbf{A}))$ for $i \in \{1, 2\}$.*

PROOF. If t is any $k+1$ -ary term for $\mathbf{A}$ and $\bar{c}, \bar{d} \in A^k$, then we put $E_1^{\mathbf{A}}(t, \bar{c}, \bar{d}) = E^{\mathbf{A}}(t, \bar{c}, \bar{d})$ as defined above, and put

$$E_2^{\mathbf{A}}(t, \bar{c}, \bar{d}) = \{a \in A : \langle t^{\mathbf{A}}(a, \bar{c}), t^{\mathbf{A}}(a, \bar{d})\rangle \in Z_1(\mathbf{A})\}.$$

Let Σ_i ($i \in \{1, 2\}$) denote the collection of all sets of the form $E_i^{\mathbf{A}}(t, \bar{c}, \bar{d})$. By definition, we have that $\langle x, y\rangle \in Z_1(\mathbf{A})$ is equivalent to: for all $S \in \Sigma_1$, $x \in S$ iff $y \in S$. It can readily be checked that $\langle x, y\rangle \in Z_2(\mathbf{A})$ is equivalent to: for all $S \in \Sigma_2$, $x \in S$ iff $y \in S$.

Since $\mathbf{A}$ is finite, each of Σ_1 and Σ_2 is finite; therefore, there is a positive integer m so that for $i \in \{1, 2\}$ we have

$$\Sigma_i = \{E_i^{\mathbf{A}}(t, \bar{c}, \bar{d}) : t \text{ is an } m+1 \text{ -ary term }, \ \bar{c}, \bar{d} \in A^m\}.$$

There are finitely many $m+1$ -ary terms $t_1, \ldots, t_k$ such that $t_1^{\mathbf{A}}, \ldots, t_k^{\mathbf{A}}$ includes all the $m+1$ -ary term operations of $\mathbf{A}$.

Now it is easily verified that the formula $\mathrm{CEN}_1(x, y)$:

$$\bigwedge_{1 \le i \le k} (\forall \bar{u}, \bar{v}) \left(t_i(x, \bar{u}) \approx t_i(x, \bar{v}) \leftrightarrow t_i(y, \bar{u}) \approx t_i(y, \bar{v})\right)$$

defines $Z_1(\mathbf{A})$ in $\mathbf{A}$, and also defines $Z_1(\mathbf{D})$ in $\mathbf{D}$ since $\mathbf{D}$ is ae-closed. One can also verify that the formula $\mathrm{CEN}_2(x, y)$:

$$\bigwedge_{1 \le i \le k} (\forall \bar{u}, \bar{v}) \left(\mathrm{CEN}_1(t_i(x, \bar{u}), t_i(x, \bar{v})) \leftrightarrow \mathrm{CEN}_1(t_i(y, \bar{u}), t_i(y, \bar{v}))\right)$$

defines $Z_2(\mathbf{A})$ in $\mathbf{A}$, and defines $Z_2(\mathbf{D})$ in $\mathbf{D}$. The assertion that $Z_i(\mathbf{D}) = D(Z_i(\mathbf{A}))$ follows from the fact that $\mathbf{D}$ is ae-closed. □

The following lemma is a key result for proving Theorem 2.1. Its proof is deferred until the end of the chapter.

LEMMA 2.4 *Let $\mathbf{A}$ be a finite algebra and U be a $Z_2(\mathbf{A})$-equivalence class containing at least two $Z_1(\mathbf{A})$-equivalence classes. Then there is a set $Q \subseteq U$ such that:*

(i) *Q is the union of two $Z_1(\mathbf{A})$-equivalence classes; and*

(ii) *$D(Q)$ is definable in every ae-closed diagonal subalgebra $\mathbf{D}$ of an algebra $\mathbf{A}^X$ (by a formula that is independent of $\mathbf{D}$ and X, and using parameters from $\mathbf{D}$).*

Proof of Theorem 2.1. Assume that $\mathbf{A}$ is a finite algebra and that $0_A < Z_1 < Z_2$ where $Z_1 = Z(\mathbf{A})$ and $Z_2/Z_1 = Z(\mathbf{A}/Z_1)$. Let $\mathcal{K}$ denote the class of ae-closed diagonal subdirect powers of $\mathbf{A}$. We are going to interpret an unstructured class of graphs into $\mathcal{K}$, which will show that $\mathcal{K}$ is unstructured and imply that $\mathbf{A} \notin \mathcal{V}$.

We begin by choosing a Z_2-equivalence class U which contains at least two Z_1-equivalence classes. By Lemma 2.3, in every $\mathbf{D} \in \mathcal{K}$, the relations $D(Z_1)$ and $D(Z_2)$ are definable, and the formulas defining them do not depend on $\mathbf{D}$. By Lemma 2.4, we can choose a pair of elements of A, call them 0 and 1, such that $\langle 0, 1\rangle \in Z_2 - Z_1$, and where

$$Q = (0/Z_1) \cup (1/Z_1),$$

there is a formula that defines the set $D(Q)$ in $\mathbf{D}$ whenever $\mathbf{D} \in \mathcal{K}$. We can switch 0 and 1 if necessary so that there exists a term $u(x, \bar{y})$ and sequences $\bar{c}, \bar{d}$ of elements such that $0 \in E^{\mathbf{A}}(u, \bar{c}, \bar{d})$ and $1 \notin E^{\mathbf{A}}(u, \bar{c}, \bar{d})$. (See Definition 2.2 and recall that $\langle 0, 1\rangle \notin Z_1$.) This fact will be used to show that the natural order relation on certain subsets $S \subseteq \{0,1\}^X \cap D$ is definable in every $\mathbf{D} \in \mathcal{K}$.

The algebra $\mathbf{R} = (\mathbf{A}|_Q)/(Z_1|_Q)$ is a two-element minimal algebra of type **1** or **2**. (See the second paragraph of §0.6 and Definition 0.29.) Our scheme of interpretation will be different depending on which of these two cases $\mathbf{R}$ falls into. We begin by considering the first case.

CASE (I) : $\mathbf{R}$ is essentially unary.

We shall be working with sets $S \subseteq \{0,1\}^X$ and with the subuniverses they generate in $\mathbf{A}^X$. Because we are in Case (I), it will be appropriate to restrict our attention to those S which are diagonal subuniverses of the algebra $\langle \{0,1\}, {}^-\rangle^X$, where ${}^-$ is the Boolean complement. For such a subuniverse S, let $\mathbf{S} = \langle S, \leq, \text{MEET}, \text{JOIN}\rangle$ be the relational structure in which $\leq$ is the ordering on S inherited from the Boolean algebra $\mathbf{2}^X$, and where $\text{MEET}(a, b, c)$ is true iff c is the meet of a and b calculated in $\mathbf{2}^X$, and $\text{JOIN}(a, b, c)$ is true iff c is the Boolean join of a and b. Finally, let $\mathcal{C}$ be the class of all such structures $\mathbf{S}$. The proof in Case (I) amounts to showing that there is a subclass of $\mathcal{C}$ which is interpretable into $\mathcal{K}$, and into which an unstructured class of graphs can be interpreted. We begin by defining this subclass of $\mathcal{C}$.

Definition 2.5 $\mathcal{C}_1$ is the class of all structures $\mathbf{S} = \langle S, \leq, \text{MEET}, \text{JOIN} \rangle$ such that $S \subseteq \{0,1\}^X$ for some set X and the following conditions hold.

(1) S contains $\hat{0}$ and $\hat{1}$ and is closed under $^-$.

(2) $\leq$, MEET and JOIN are derived from the Boolean algebra $\mathbf{2}^X$ as described above.

(3) $|S| > 2$.

(4) $\mathbf{S}$ has no infinite chains.

(5) If $a \prec b \prec c$ in $\mathbf{S}$, then the relative complement of b in $I[a,c]$ calculated in $\mathbf{2}^X$ does not belong to S.

LEMMA 2.6 *$\mathcal{C}_1$ is interpretable into $\mathcal{K}$ (assuming Case (I)).*

PROOF. Given $\mathbf{S} \in \mathcal{C}_1$, say $S \subseteq \{0,1\}^X$, let $T = X \cup \{0,1\}^S$, and let $X^* = \bigcup\{X_t : t \in T\}$ where every set X_t is infinite and $X_t \cap X_{t'} = \emptyset$ when $t \neq t'$. (We can assume that T is the disjoint union of X and $\{0,1\}^S$; a small but crucial fact if our argument is to succeed.)

The algebra we are constructing will be defined as the ae-closure of the subalgebra of $\mathbf{A}^{X^*}$ generated by a certain set of functions. These functions are constant on each one of the sets X_t. For $s \in S$ we define a function $s^* \in A^{X^*}$, essentially by "extending" s to X^*. (Recall that $S \subseteq \{0,1\}^X \subseteq A^X$. For $t \in T$ we put

$$s^*(X_t) = \begin{cases} \{s(t)\} & \text{if } t \in X\,, \\ \{t(s)\} & \text{if } t \in T - X\,. \end{cases} \tag{2.1}$$

We define one further function μ.

$$\mu(X_t) = \begin{cases} \{1\} & \text{if } t \in X \\ \{0\} & \text{if } t \in T - X\,. \end{cases} \tag{2.2}$$

Let B^* be the collection consisting of μ and the s^* $(s \in S)$. Let $\mathbf{D}'$ be the diagonal subalgebra of $\mathbf{A}^{X^*}$ generated by B^*, and let $\mathbf{D}$ be the ae-closure of $\mathbf{D}'$. We shall interpret $\mathbf{S}$ into $\mathbf{D}$.

Before proceeding further, note that $\langle B^*, \leq \rangle$ is isomorphic in a natural way to a sub-ordered set $\langle B, \leq \rangle$ of $\langle \{0,1\}, \leq \rangle^T$. We need to work with $B^* \subseteq A^{X^*}$ rather than B in order to avoid having the closure under ae-equivalence destroy the structure that we are trying to define in $\mathbf{D}$. Ultimately, we will need to recover B from B^*; that is, we will need to show that the relation $\overset{\mathrm{ae}}{=}$ (restricted to some definable subset of $\mathbf{D}$ containing B^*) is definable in $\mathbf{D}$. The replacement of X by X^*, and the extension of the functions in S to X^*, is the trick which will make this possible.

We now begin the work of constructing an interpretation. As usual, Σ denotes the finite set of all subsets of A of the form $E^{\mathbf{A}}(t, \bar{c}, \bar{d})$, with t ranging over all terms in the language of $\mathbf{A}$. We now choose a set $W \in \Sigma$ which contains 0 but not 1, and which is maximal with respect to possessing this property. We can suppose that $W = E^{\mathbf{A}}(u, \bar{c}_0, \bar{d}_0)$ where $\bar{c}_0, \bar{d}_0 \in A^m$ and u is an $m+1$ -ary term. We denote by $\bar{c}'$ and $\bar{d}'$ the corresponding m-tuples of constant elements of D. For $\bar{\alpha}, \bar{\beta} \in D^m$ we write $E(\bar{\alpha}, \bar{\beta})$ for the set $E^{\mathbf{D}}(u, \bar{\alpha}, \bar{\beta})$ (which is a subset of D). We now define some relations on, and a subset of, $(D^m)^2$.

$$
\begin{aligned}
(\bar{\alpha}, \bar{\beta}) \leq (\bar{\gamma}, \bar{\delta}) \quad &\text{iff} \quad E(\bar{\alpha}, \bar{\beta}) \subseteq E(\bar{\gamma}, \bar{\delta}). \\
(\bar{\alpha}, \bar{\beta}) < (\bar{\gamma}, \bar{\delta}) \quad &\text{iff} \quad E(\bar{\alpha}, \bar{\beta}) \text{ is a proper subset of } E(\bar{\gamma}, \bar{\delta}). \\
(\bar{\alpha}, \bar{\beta}) \sim (\bar{\gamma}, \bar{\delta}) \quad &\text{iff} \quad E(\bar{\alpha}, \bar{\beta}) = E(\bar{\gamma}, \bar{\delta}). \qquad (2.3) \\
(\bar{\alpha}, \bar{\beta}) \in P \quad &\text{iff} \quad (\bar{c}', \bar{d}') \leq (\bar{\alpha}, \bar{\beta}) \text{ and } \hat{1} \notin E(\bar{\alpha}, \bar{\beta}) \text{ and} \\
&\qquad (\forall \bar{\gamma}, \bar{\delta})[(\bar{\alpha}, \bar{\beta}) < (\bar{\gamma}, \bar{\delta}) \rightarrow \hat{1} \in E(\bar{\gamma}, \bar{\delta})].
\end{aligned}
$$

The next claim follows easily from the fact that $\mathbf{D}$ is diagonal and ae-closed, and by our choice of $(\bar{c}', \bar{d}')$.

Claim 1. We have $(\bar{\alpha}, \bar{\beta}) \in P$ iff there exists $x \in X^*$ such that $E(\bar{\alpha}, \bar{\beta}) = \{\delta \in D : \delta(x) \in W\}$. This element $\chi(\bar{\alpha}, \bar{\beta}) = x$ corresponding to $(\bar{\alpha}, \bar{\beta}) \in P$ is unique, and we have a mapping χ of P onto X^* such that $\chi(\bar{\alpha}, \bar{\beta}) = \chi(\bar{\gamma}, \bar{\delta})$ iff $(\bar{\alpha}, \bar{\beta}) \sim (\bar{\gamma}, \bar{\delta})$.

We note that if $\bar{\alpha}(x) = \bar{\beta}(x)$ for all $x \in X^* - \{x_0\}$ and $(\bar{\alpha}(x_0), \bar{\beta}(x_0)) = (\bar{c}_0, \bar{d}_0)$, then $(\bar{\alpha}, \bar{\beta}) \in P$ and $\chi(\bar{\alpha}, \bar{\beta}) = x_0$.

The set P is equivalent to a $2m$-ary relation on D that is definable using the constant functions as parameters (by (2.3)). Likewise, the relations $\leq$, $<$, and $\sim$, and their restrictions to P are definable. We regard the elements of P as codes for the points of X^*.

We define

$$Y^* = \bigcup \{X_t : t \in X\} \ (= \mu^{-1}(1)). \qquad (2.4)$$

Note that the set $W = E^{\mathbf{A}}(u, \bar{c}_0, \bar{d}_0)$ contains $0/Z_1$ and is disjoint from $1/Z_1$; hence if $\alpha \in D(Q)$, $(\bar{\gamma}, \bar{\delta}) \in P$, and $\chi(\bar{\gamma}, \bar{\delta}) = x$ then

$$\alpha(x) Z_1 1 \leftrightarrow \alpha \notin E(\bar{\gamma}, \bar{\delta}); \qquad \text{and thus}$$

$$\chi(\bar{\gamma}, \bar{\delta}) \in Y^* \leftrightarrow \mu \notin E(\bar{\gamma}, \bar{\delta}). \qquad (2.5)$$

It follows from (2.5) that for $\alpha, \beta \in D(Q)$ the condition $[\![\alpha Z_1 \beta]\!] \supseteq Y^*$ is equivalent to the satisfaction of the formula $\mathrm{Z}^\mu(\alpha,\beta)$:

$$(\forall(\bar{\gamma},\bar{\delta}) \in P)[\mu \notin E(\bar{\gamma},\bar{\delta}) \rightarrow (\alpha \in E(\bar{\gamma},\bar{\delta}) \leftrightarrow \beta \in E(\bar{\gamma},\bar{\delta}))].$$

Obviously, we have a mapping $F : D(Q) \rightarrow \{0,1\}^{Y^*}$ defined by $F(\alpha) = k\alpha|_{Y^*}$ where $k : Q \rightarrow \{0,1\}$ is the map which collapses $0/Z_1$ to 0 and $1/Z_1$ to 1. The kernel of F is the relation on $D(Q)$ defined by the formula $\mathrm{Z}^\mu(x,y)$.

We need to study the structure

$$\mathbf{L} = \langle L, \leq, \text{MEET}, \text{JOIN} \rangle,$$

where $L = F(D(Q))$ (the image of $D(Q)$ under F). This is the structure belonging to $\mathcal{C}$ that is determined by the inclusion $L \subseteq \{0,1\}^{Y^*}$. We can easily see that the F-inverse images of $\leq$, and of MEET and JOIN, on L are definable in $\mathbf{D}$ in the following way.

$$\text{MEET}(F(\alpha),F(\beta),F(\rho)) \text{ iff } (\forall(\bar{\gamma},\bar{\delta}) \in P)[\mu \notin E(\bar{\gamma},\bar{\delta}) \rightarrow \{\rho \notin E(\bar{\gamma},\bar{\delta}) \leftrightarrow (\{\alpha,\beta\} \cap E(\bar{\gamma},\bar{\delta}) = \emptyset)\}]$$

$$\text{JOIN}(F(\alpha),F(\beta),F(\rho)) \text{ iff } (\forall(\bar{\gamma},\bar{\delta}) \in P)[\mu \notin E(\bar{\gamma},\bar{\delta}) \rightarrow \{\rho \in E(\bar{\gamma},\bar{\delta}) \leftrightarrow (\{\alpha,\beta\} \subseteq E(\bar{\gamma},\bar{\delta}))\}]$$

$$F(\alpha) \leq F(\beta) \text{ iff } \text{MEET}(F(\alpha),F(\beta),F(\alpha)).$$

The equivalences above exhibit a scheme that interprets $\mathbf{L}$ into $\mathbf{D}$. More precisely, suppose:

(1) $\mathrm{Un}(x,\bar{y})$ is the formula given by Lemma 2.4 which defines $D(Q)$, for some appropriate choice of parameters from D.

(2) $\mathrm{Eq}(x_1,x_2,\bar{u},\bar{v},y,z)$ is a formula which, upon substituting $\bar{c}'$, $\bar{d}'$, $\hat{1}$ and μ for $\bar{u}$, $\bar{v}$, y and z respectively, is equivalent to $\mathrm{Z}^\mu(x_1,x_2)$.

(3) The formulas

$$\mathrm{S}_\wedge(x_1,x_2,x_3,\bar{u},\bar{v},y,z), \quad \mathrm{S}_\vee(x_1,x_2,x_3,\bar{u},\bar{v},y,z)$$

and

$$\mathrm{S}_\leq(x_1,x_2,\bar{u},\bar{v},y,z)$$

are equivalent, upon substituting $\bar{c}'$, $\bar{d}'$, $\hat{1}$ and μ for $\bar{u}$, $\bar{v}$, y and z respectively, to the formulas on the right sides of the above displayed bi-implications (with free variables x_1, x_2, x_3 replacing α, β, ρ).

Then $\bar{\psi} = \langle \text{Un}, \text{Eq}, \text{S}_{\leq}, \text{S}_{\wedge}, \text{S}_{\vee} \rangle$ is a scheme which we have proved interprets **L** into **D**. Note that this scheme is independent of any variation in the choice of the structure $\mathbf{S} \in \mathcal{C}_1$. Thus to prove this lemma, it remains to interpret **S** into **L**, again by a scheme that is independent of the original **S**.

Notice that for each of the generating elements s^*, we have $F(s^*) = s^*|_{Y^*}$. We denote this element $F(s^*) \in L$ by λ_s, and we put

$$L' = \{\lambda_s : s \in S\}. \tag{2.6}$$

From (2.1) we have that $\lambda_s(y) = s(t)$ when $y \in X_t$; and it readily follows that the map $s \mapsto \lambda_s$ is an isomorphism of **S** onto the structure

$$\mathbf{L}' = \langle L', \leq, \text{MEET}, \text{JOIN} \rangle$$

belonging to $\mathcal{C}$.

Claim 2. L is the ae-closure of $L' = \{\lambda_s : s \in S\}$.

To prove this claim, suppose that $\alpha \in D(Q)$. We should like to prove that $F(\alpha)$ is almost equal to a member of L'. Now we have $\alpha \overset{\text{ae}}{=} \alpha'$ for some $\alpha' \in D'$, and since each member of D' takes each of its values infinitely often, we have that $\alpha' \in D(Q)$. It will suffice to show that $F(\alpha')$ belongs to L'.

From the definition of $\mathbf{D}'$, there exists a polynomial operation f of **A** such that $f^{(X)}$ (f acting coordinatewise) applied to some of the generators s^*, μ of $\mathbf{D}'$, gives α'. Thus we can write

$$f^{(X)}(\mu, s_0^*, \ldots, s_{k-1}^*) = \alpha', \tag{2.7}$$

where $s_0, \ldots, s_{k-1}$ are distinct elements of S, and f is a polynomial of **A**. We can assume that $F(\alpha')$ is not constant, (i.e., α' is not constant on Y^*, modulo Z_1) since otherwise the desired result follows from the fact that $\hat{0}, \hat{1} \in S$.

From our definition of T, and of the generating functions of $\mathbf{D}'$, for every $\sigma \in \{0,1\}^k$ there is a corresponding $t \in T - X$ and $x \in X_t$ such that

$$\langle \mu(x), s_0^*(x), \ldots, s_{k-1}^*(x) \rangle = \langle 0, \sigma(0), \ldots, \sigma(k-1) \rangle.$$

Thus by (2.7) and the fact that all values of α' lie in Q, the polynomial $h(\bar{y}) = f(0, \bar{y})$ maps $\{0,1\}^k$ into Q. Since $Q = 0/Z_1 \cup 1/Z_1$ and Z_1 is a congruence, we can deduce that Q is closed under h. Thus for $\bar{y} \in Q^k$ there are only two possible values, modulo Z_1, of $f(0, \bar{y})$. Notice that since $\langle 0, 1 \rangle \in Z_2$, for all $\bar{y}, \bar{z}$ we have

$$f(0, \bar{y}) \, Z_1 \, f(0, \bar{z}) \leftrightarrow f(1, \bar{y}) \, Z_1 \, f(1, \bar{z}).$$

Thus it follows that $f(1,\bar{y})$ $(\bar{y} \in Q^k)$ takes on at most two values modulo Z_1. Now for $x \in Y^*$ we have

$$f(1, s_0^*(x), \ldots, s_{k-1}^*(x)) = \alpha'(x) \in Q$$

by (2.7). Since $\alpha'(x)$ is not constant on Y^* modulo Z_1, it follows that the two Z_1-equivalence classes that contain the values of $f(1,\bar{y})$ $(\bar{y} \in Q^k)$ can be none other than $0/Z_1$ and $1/Z_1$; i.e., we have that Q is closed under the polynomial $f(1,\bar{y})$.

Hence $f(1,\bar{y})$, restricted to Q, induces an operation of $\mathbf{R}$, which must be essentially unary. Since $F(\alpha')$ is not constant, this operation must depend on precisely one variable, say the ith. This leads directly to the conclusion that $F(\alpha')$ is either λ_s or $\lambda_s{}^-$ where $s = s_i$. Since $s \in S$ implies $s^- \in S$, the proof of Claim 2 is finished.

The proof of Lemma 2.6 will now be completed in two steps. First we show that if the relation $\stackrel{\text{ae}}{\equiv}$ in L is definable in $\mathbf{L}$, then $\mathbf{S}$ is interpretable in $\mathbf{L}$. Then, to conclude the argument, we prove that $\stackrel{\text{ae}}{\equiv}$ is definable in $\mathbf{L}$.

Assuming that $\stackrel{\text{ae}}{\equiv}$ is definable in $\mathbf{L}$, we let H be the map from L onto S such that $H(\alpha)$ is the unique $s \in S$ for which $\lambda_s \stackrel{\text{ae}}{\equiv} \alpha$. That H is well-defined follows from Claim 2, and from the fact that each function in L' is constant on each infinite set X_t, so that two distinct members of L' cannot be ae-equivalent. Most of the claims made in this paragraph also rely on this fact. We see that the kernel of H is the definable relation $\stackrel{\text{ae}}{\equiv}$. The H-inverse images of the fundamental relations of $\mathbf{S}$ are definable in $\mathbf{L}$ as well. For example, it is obvious that $H(\alpha) \leq H(\beta)$ iff there exists $\alpha' \stackrel{\text{ae}}{\equiv} \alpha$ such that $\alpha' \leq \beta$. If $H(\alpha_i) = s_i$ $(i \in \{0,1,2\})$ then $\mathbf{S} \models \text{MEET}(s_0, s_1, s_2)$ iff λ_{s_2} is the Boolean meet of λ_{s_0} and λ_{s_1} iff there exists $\alpha' \stackrel{\text{ae}}{\equiv} \alpha_2$ in L such that $\mathbf{L} \models \text{MEET}(\alpha_0, \alpha_1, \alpha')$. The relation

$$\{\langle \alpha_0, \alpha_1, \alpha_2 \rangle \in L^3 : \mathbf{S} \models \text{JOIN}(H(\alpha_0), H(\alpha_1), H(\alpha_2))\}$$

is defined in $\mathbf{L}$ by an analogous formula. Thus we have an interpretation of $\mathbf{S}$ in $\mathbf{L}$. The scheme of formulas used in this interpretation, like that used in the earlier interpretation of $\mathbf{L}$ in $\mathbf{D}$, is independent of the choice of $\mathbf{S} \in \mathcal{C}_1$. Thus, combining the two interpretations gives an interpretation of $\mathcal{C}_1$ in $\mathcal{K}$.

To finish the proof of Lemma 2.6, we now show that the relation of ae-equivalence is first-order definable in $\mathbf{L}$. The key is to consider an auxiliary definable relation; for $\alpha, \beta \in L$ we write $\alpha \circ \beta$ iff there exist $\gamma, \delta \in L$ such that $\gamma \leq \alpha, \beta \leq \delta$ and the interval $I[\gamma, \delta]$ in $\langle L, \leq \rangle$ is closed under the Boolean operations of meet, join, and relative complementation in $\mathbf{2}^{Y^*}$. This relation $\circ$ is definable using the ternary relations MEET and JOIN of $\mathbf{L}$; it is a reflexive and symmetric relation on L, but is not an equivalence relation. We make two claims.

Claim 3. If $\alpha \circ \beta$, $\alpha' \stackrel{\mathrm{ae}}{=} \alpha$, and $\beta' \stackrel{\mathrm{ae}}{=} \beta$, then $\alpha' \circ \beta'$.

Claim 4. If $\alpha, \beta \in L'$ then $\mathbf{L} \models \alpha \circ \beta$ iff $\mathbf{L}' \models \alpha \circ \beta$.

The proofs of these assertions are straightforward, using just the facts that L is the $\stackrel{\mathrm{ae}}{=}$ -closure of L', and Y^* is partitioned into infinite sets on which each member of L' is constant.

Note that Claim 4 translates, via the isomorphism of $\mathbf{S}$ with $\mathbf{L}'$, into the statement

$$\text{if } s, s' \in S \text{ then } \mathbf{S} \models s \circ s' \text{ iff } \mathbf{L} \models \lambda_s \circ \lambda_{s'}.$$

Next we observe that the conditions defining $\mathcal{C}_1$, especially condition (5), imply that when $s_0 < s_1$ in $\mathbf{S}$ then $\mathbf{S} \models s_0 \circ s_1$ iff s_1 covers s_0 in $\mathbf{S}$. This means that for any $s_0, s_1 \in S$ we have $\mathbf{S} \models s_0 \circ s_1$ iff $s_0 \prec s_1$ or $s_1 \prec s_0$ or $s_0 = s_1$. The definition of $\mathcal{C}_1$ also implies that when $s_0 \prec s_1$ in $\mathbf{S}$, then there is an element $s \in S$ such that either $s \prec s_0$ or $s_1 \prec s$. These facts should make the next assertion obvious.

$$\mathbf{L}' \models (\forall x, y)(x \approx y \leftrightarrow (\forall z)(z \circ \alpha \leftrightarrow z \circ \beta)).$$

In view of Claims 3 and 4, this yields

$$\text{for } \alpha, \beta \in L,\ \alpha \stackrel{\mathrm{ae}}{=} \beta \text{ iff } \mathbf{L} \models (\forall z)(z \circ \alpha \leftrightarrow z \circ \beta).$$

Thus $\stackrel{\mathrm{ae}}{=}$ is a definable binary relation in $\mathbf{L}$. □

LEMMA 2.7 *The class $\mathcal{C}_1$ is unstructured.*

PROOF. It is very easy to find an interpretation of the class of all graphs into the class of graphs that have at least four vertices, and have the property that for each vertex p there is a vertex q such that $\{p, q\}$ is not an edge. We shall interpret this latter class into $\mathcal{C}_1$. Let $\mathbf{G} = \langle P, E \rangle$ be a graph of this class. We can harmlessly assume that $P \cap E = \emptyset$, and we put $X = P \cup E$. For each $p \in P$ define $\chi_p \in \{0,1\}^X$ by

$$\chi_p(x) = \begin{cases} 1 & \text{if } x = p \text{ or } p \in x \in E \\ 0 & \text{otherwise}. \end{cases}$$

Let $S \subseteq \{0,1\}^X$ be the set of all the functions χ_p, their complements $\chi_p{}^-$, and the two constant functions $\hat{0}$ and $\hat{1}$.

It is easy to visualize the structure of the ordered set $\langle S, \leq \rangle$, and then to prove that it belongs to $\mathcal{C}_1$. All the elements χ_p cover $\hat{0}$. Their complements

$\chi_p{}^-$ are covered by $\hat{1}$. We have $\chi_p \leq \chi_q{}^-$ iff $\chi_p < \chi_q{}^-$ iff $p \neq q$ and $\{p,q\} \notin E$; hence each atom χ_p is covered by some co-atom $\chi_q{}^-$, and each $\chi_q{}^-$ covers some atom χ_p. All maximal chains in $\langle S, \leq \rangle$ have four elements. Since the three-element covering chains in $\langle S, \leq \rangle$ are thus of two sorts, namely $\hat{0} \prec \chi_p \prec \chi_q{}^-$ and $\chi_p \prec \chi_q{}^- \prec \hat{1}$, it is easy to check that when $a \prec b \prec c$, then the Boolean relative complement of b in the interval $I[a, c]$ does not belong to S. (The existence of at least four distinct vertices is needed at this point.) Thus $\mathbf{S} = \langle S, \leq, \text{MEET}, \text{JOIN} \rangle$ belongs to $\mathcal{C}_1$.

It is very easy to interpret $\mathbf{G}$ in $\mathbf{S}$. The elements $p \in P$ are identified with the atoms χ_p of $\mathbf{S}$; and it is easily established that $\{p,q\} \in E$ is equivalent to $\mathbf{S} \models \neg\chi_p = \chi_q \wedge \neg\text{MEET}(\chi_p, \chi_q, \hat{0})$. □

Lemmas 2.6 and 2.7 imply that $\mathcal{K}$ is unstructured in Case (I). Now we consider the other case.

CASE (II): $\mathbf{R}$ is polynomially equivalent with a two-element group.

Because we are in Case (II), it will be appropriate to consider those sets $S \subseteq \{0,1\}^X$ that are diagonal subuniverses of $\langle \{0,1\}, + \rangle^X$, where $+$ is Boolean addition (or symmetric difference). We now define the class $\mathcal{C}_2$ of models which we shall interpret into $\mathcal{K}$ in Case (II). The proof that $\mathcal{C}_2$ is unstructured will be given immediately following the proof that this class interprets into $\mathcal{K}$.

Definition 2.8 $\mathcal{C}_2$ is the class of all models $\langle S, +, \leq, \text{MEET}, \text{JOIN} \rangle$ such that $\langle S, + \rangle$ is a subgroup of the Boolean group $\langle \{0,1\}, + \rangle^X$ for some set X, and S has subsets S_0 and S_1 and elements ψ_0, ψ_1, ψ_e satisfying the following conditions.

(1) $\langle S, \leq, \text{MEET}, \text{JOIN} \rangle$ is a member of $\mathcal{C}$. (See the paragraph preceding Definition 2.5.)

(2) Where $W_i = \psi_i^{-1}(1)$ $(i \in \{0,1\})$ and $E = \psi_e^{-1}(1)$, the sets W_0, W_1 and E partition X and $|W_i| \geq 2$.

(3) S_i $(i \in \{0,1\})$ is a nonempty subset of the interval $I[0, \psi_i + \psi_e]$ and is closed under Boolean meet, join and relative complement in this interval. Moreover, $|S_i| > 2$.

(4) There exist surjective maps $\pi_i : E \to W_i$ $(i \in \{0,1\})$ such that for all $s \in S_i$ and all $e \in E$, $s(e) = s(\pi_i(e))$.

(5) $S = S_0 + S_1 + \{\hat{0}, \psi_e\}$; i.e., for every $s \in S$ there are $s_0 \in S_0$ and $s_1 \in S_1$ and $\psi \in \{\hat{0}, \psi_e\}$ such that $s = s_0 + s_1 + \psi$.

LEMMA 2.9 *$\mathcal{C}_2$ is interpretable into $\mathcal{K}$ (assuming Case (II)).*

PROOF. Given $\mathbf{S} \in \mathcal{C}_2$, $S \subseteq \{0,1\}^X$, we proceed exactly as in the proof of Lemma 2.6, up to the point where the structure $\mathbf{L}$ in $\mathcal{C}$, $L = F(D(Q))$, is introduced and shown to be interpretable in $\mathbf{D}$. Since we are in Case (II), there is a polynomial operation $\text{Sum}(x,y)$ of $\mathbf{A}$, under which Q is closed, which induces the operation of an Abelian group on Q/Z_1 whose zero element is $0/Z_1$. The map k from Q to $\{0,1\}$ which takes all elements of $0/Z_1$ to 0 and all elements of $1/Z_1$ to 1 is a homomorphism of $\langle Q, \text{Sum}|_Q \rangle$ onto $\langle \{0,1\}, + \rangle$. Since $\text{Sum}(x,y)$ acting coordinatewise in A^X is a polynomial operation of $\mathbf{D}$, it follows that $L = F(D(Q))$ is a subgroup of $\langle \{0,1\}, + \rangle^X$. We can extend our interpretation of $\mathbf{L}$ in $\mathbf{D}$ to include the operation $+$, since "$H(\alpha)+H(\beta) = H(\delta)$" is equivalent to "$\text{Sum}(\alpha,\beta)\, Z_1\, \delta$", so long as $\{\alpha,\beta,\delta\} \subseteq D(Q)$. Thus, changing notation for the remainder of the proof, we have

$$\mathbf{L} = \langle L, +, \leq, \text{MEET}, \text{JOIN} \rangle.$$

It remains for us to interpret $\mathbf{S}$ into $\mathbf{L}$. The next step is to prove Claim 2 in Case (II). We define $\lambda_s = s^*|_{Y^*}$ and $L' = \{\lambda_s : s \in S\}$ just as before.

Claim 2. L is the ae-closure of the set L'.

The proof is the same as the proof of Claim 2 in Lemma 2.6, up to the point where it is shown that the polynomial $f(1,\bar{y})$, restricted to Q, induces an operation of $\mathbf{R}$. Since we are in Case (II), this means that $F(\alpha')$ is of the form $\lambda_{s_0} + \cdots + \lambda_{s_{k-1}} + \varepsilon$ with $\varepsilon \in \{\hat{0}, \hat{1}\}$. Note that Definition 2.8 implies that $\hat{1}$ $(= \psi_0 + \psi_1 + \psi_e)$ is in S. Thus $F(\alpha')$ is of the form λ_s for some $s \in S$, since S is closed under $+$. The proof of Claim 2 is complete.

From Claim 2, we have a function H from L onto S defined by taking $H(\alpha)$ to be the unique $s \in S$ for which $\lambda_s \stackrel{\text{ae}}{=} \alpha$. Just as before, the kernel of H is $\stackrel{\text{ae}}{=}$; and the H-inverse image of each of the fundamental operations and relations of $\mathbf{S}$ can easily be seen to be definable in $\mathbf{L}$, provided that $\stackrel{\text{ae}}{=}$ is definable.

To establish the definability of $\stackrel{\text{ae}}{=}$ in $\mathbf{L}$, we use the properties of $\mathbf{S}$ laid down in Definition 2.8. Let ψ_0, ψ_1 and ψ_e be the three elements of S, and S_0 and S_1 be the subsets of S, that satisfy the conditions of the definition. Then let $\tau_i = \lambda_{\psi_i}$ (for $i \in \{0,1\}$) and let $\tau_e = \lambda_{\psi_e}$.

Claim 3. The only elements χ of L' such that $\chi \leq \tau_0$ are $\hat{0}$ and τ_0. Similar statements hold for τ_1 and τ_e.

The validity of these assertions is a consequence of the isomorphism of $\mathbf{L}'$ with $\mathbf{S}$, and of the properties assumed for $\mathbf{S}$. Suppose that $s \in S$ and

$s \le \psi_0$. Write $s = s_0 + s_1 + \varepsilon$ with $s_i \le \psi_i + \psi_e$ and $\varepsilon \in \{\hat{0}, \psi_e\}$ (using 2.8 (5, 3)). Then it follows, by 2.8 (2, 4), that if $s_1 \ne \hat{0}$ then $s_1(x) = 1$ for some $x \in W_1$; and this means that $s(x) = 1$, since $s_0(x) = \varepsilon(x) = 0$. But since $s \le \psi_0$, these considerations imply that $s_1 = \hat{0}$. If $s_0 = \hat{0}$, then $s = \hat{0}$ since $\psi_e \not\le \psi_0$. However, if $s_0 \ne \hat{0}$, then $\varepsilon = \psi_e$ since 2.8 (4) imply that $s_0 \not\le \psi_0$. Thus if $s \ne \hat{0}$ then $s = s_0 + \psi_e$. Then since $s(x) = 0$ and $\psi_e(x) = 1$ for $x \in E$, it follows that $E \subseteq s_0^{-1}(1)$. Then 2.8 (4) imply that $s_0 = \psi_0 + \psi_e$, and hence $s = \psi_0$. So we have proved that $\hat{0}$ and ψ_0 are the only elements of S below ψ_0. The same facts hold for ψ_1 and ψ_e, and they imply Claim 3 by virtue of the isomorphism of $\mathbf{L}'$ with $\mathbf{S}$.

Claim 4. Suppose that $\tau \in \{\tau_0, \tau_1, \tau_e\}$ and $\chi \in L$ satisfies $\chi \le \tau$. Then $\chi \overset{\text{ae}}{=} \hat{0}$ iff for every $\alpha \in L$ there exists $\delta \in L$ such that $\mathbf{L} \models \text{MEET}(\chi, \alpha, \delta)$.

We shall prove this claim for $\tau = \tau_1$. The proof for $\tau = \tau_0$ is the same, and the proof for $\tau = \tau_e$ is easier. So let $\chi \le \tau_1$. Then by Claims 2 and 3, we have $\chi \overset{\text{ae}}{=} \hat{0}$ or $\chi \overset{\text{ae}}{=} \tau_1$. If $\chi \overset{\text{ae}}{=} \hat{0}$, then the Boolean meet of χ with any member of L belongs to L, since L is $\overset{\text{ae}}{=}$-closed. Now suppose that $\chi \overset{\text{ae}}{=} \tau_1$. Choose, by 2.8 (3), $s \in S_1$ such that $\hat{0} < s < \psi_1 + \psi_e$. Assuming that $\delta \in L$ and $\mathbf{L} \models \text{MEET}(\chi, \lambda_s, \delta)$, we shall obtain a contradiction, and that will finish the proof of this claim.

Let $s' \in S$ be the element such that $\delta \overset{\text{ae}}{=} \lambda_{s'}$. Since also $\chi \overset{\text{ae}}{=} \tau_1$, and τ_1, λ_s, $\lambda_{s'}$ are constant on each infinite set X_t, it follows that $\lambda_{s'}$ is the Boolean meet of τ_1 and λ_s; thus s' is the Boolean meet of ψ_1 and s. By 2.8 (4), $s < \psi_1 + \psi_e$ and $s \in S_1$ imply that $s' < \psi_1$. Thus, by Claim 3, $s' = \hat{0}$. On the other hand, $\hat{0} < s \in S_1$ implies, by 2.8 (4), that s is non-zero at some $x \in W_1$, which means that s' cannot be $\hat{0}$. So Claim 4 is proved.

Now if $\alpha, \beta \in L$ then $\alpha \overset{\text{ae}}{=} \beta$ iff $\chi \overset{\text{ae}}{=} \hat{0}$ where $\chi = \alpha + \beta$; and $\chi \overset{\text{ae}}{=} \hat{0}$ iff $\chi = \chi_0 + \chi_1 + \chi_e$ where $\hat{0} \overset{\text{ae}}{=} \chi_i \le \tau_i$ ($i \in \{0, 1\}$) and $\hat{0} \overset{\text{ae}}{=} \chi_e \le \tau_e$. By Claim 4, we have that the relation "$\alpha \overset{\text{ae}}{=} \beta$" is definable in $\mathbf{L}$. This finishes our proof of Lemma 2.9. □

LEMMA 2.10 *The class $\mathcal{C}_2$ is unstructured.*

PROOF. We consider a special class of bi-partite directed graphs $\mathbf{P} = \langle P, E \rangle$; namely, those in which P is the disjoint union of subsets W_0 and W_1 such that $E \subseteq W_0 \times W_1$, and such that each of W_0 and W_1 has at least two elements, and $\pi_i(E) = W_i$ (for $i \in \{0, 1\}$) where π_i is the coordinate projection. Let $\mathcal{BG}$ denote the class of all such structures. We shall interpret $\mathcal{BG}$ into $\mathcal{C}_2$; but first, we will show that $\mathcal{BG}$ interprets the class of all graphs.

Let $\mathbf{G} = \langle V, B \rangle$ be any graph. (According to our official definition of "graph", this just means that B is a set of two-element subsets of the nonvoid set V.) We can assume that $V \cap B = \emptyset$. Choose four elements c_0, c_1, d_0, d_1 that do not belong to $V \cup B$; and put $W_0 = V \cup \{c_0, c_1\}$, $W_1 = B \cup \{d_0, d_1\}$ and $P = W_0 \cup W_1$. Then define

$$E = (V \times \{d_0, d_1\}) \cup (\{c_0, c_1\} \times B) \cup E' \text{ where}$$

$$E' = \{\langle v, b\rangle \in V \times B : v \in b\}.$$

The structure $\mathbf{P} = \langle P, E\rangle$ belongs to $\mathcal{BG}$; and it is straightforward to find an interpretation of $\mathbf{G}$ in $\mathbf{P}$.

Now to see that $\mathcal{C}_2$ interprets $\mathcal{BG}$, suppose that $\mathbf{P} = \langle P, E\rangle$ is an arbitrary member of $\mathcal{BG}$. Letting W_0, W_1 be the two subsets of P which are domain and range of E, we put $X = W_0 \cup W_1 \cup E$. We define S to consist of all functions $f \in \{0,1\}^X$ such that

$$f = f_0 \cup f_1 \cup f_e, \quad \text{where } f_i \in \{0,1\}^{W_i} \text{ and } f_e \in \{0,1\}^E;$$

and each of the functions f_0, f_1 is ae-equivalent either to $\hat{0}$ or $\hat{1}$ on its domain; and

$$\text{either } f_e = f_0\pi_0 + f_1\pi_1 \text{ or } f_e = f_0\pi_0 + f_1\pi_1 + \hat{1}.$$

We define $S_i \subseteq S$ ($i \in \{0,1\}$) to be the set of all f as above such that $f_{1-i} = \hat{0}$ and $f_e = f_i\pi_i$. We define ψ_0, ψ_1, $\psi_e \in S$ to be the characteristic functions of the sets W_0, W_1 and E.

It is not difficult to verify that S is closed under Boolean addition in $\{0,1\}^X$; and the structure $\mathbf{S} = \langle S, +, \leq, \text{MEET}, \text{JOIN}\rangle$ belongs to $\mathcal{C}_2$ (see Definition 2.8). Note that the restriction map of S_i into $\{0,1\}^{W_i}$ maps $\langle S_i, \leq\rangle$ isomorphically onto the Boolean lattice consisting of all members of $\mathbf{2}^{W_i}$ having finite or co-finite support. For each element $w \in W_i$ ($i \in \{0,1\}$) define $s_w \in S_i$ to be the atom of the Boolean lattice $\langle S_i, \leq\rangle$ that corresponds to w. (This means that $s_w(x) = 1$ iff either $x = w$, or $x \in E$ and $\pi_i(x) = w$.)

In order to build an interpretation of $\mathbf{P}$ in $\mathbf{S}$, let $A_i = \{s_w : w \in W_i\}$ for $i \in \{0,1\}$, and put $A = A_0 \cup A_1$. Choose elements $w_0 \in W_0$ and $w_1 \in W_1$. Now observe that $s \in S$ satisfies $s \in S_i$ iff $s \leq \psi_i + \psi_e$ and $\mathbf{S} \models \text{MEET}(s, s_{w_i}, d)$ for some $d \in S$. (The verification, which we omit, is straightforward. It uses the fact that each $w \in W_i$ belongs to at least one, but not to all, edges.) Therefore, S_0 and S_1 are definable in $\mathbf{S}$. Since A_i is the set of all atoms of $\langle S_i, \leq\rangle$, the sets A_0, A_1 and A are likewise definable. Let h be the bijective mapping of A onto P defined by $h(s_w) = w$.

To see that we have an interpretation of $\mathbf{P}$ in $\mathbf{S}$, all that remains to be shown is that the relation

$$h^{-1}(E) = \{\langle s, s'\rangle : \langle h(s), h(s')\rangle \in E\}$$

is definable in $\mathbf{S}$. We have $\langle s, s' \rangle \in h^{-1}(E)$ iff $s \in A_0$, $s' \in A_1$, and $\mathbf{S} \models \neg\mathrm{MEET}(s, s', \hat{0})$. □

Our proof of Theorem 2.1 would now be complete, except that we have relied on Lemma 2.4, which remains to be proved. To prove Lemma 2.4, we require the next lemma.

LEMMA 2.11 *Let U be a nonvoid subset of a finite set A, let Σ be a set of subsets of A, and let ζ be the equivalence relation over U defined by $\langle x, y \rangle \in \zeta \leftrightarrow (\forall S \in \Sigma)(x \in S \leftrightarrow y \in S)$. Suppose that $\zeta \neq U^2$. Then there exists a pair $\langle a, b \rangle \in U^2 - \zeta$ such that*

$$(\forall x \in U)[\,\{x\zeta a \vee x\zeta b\} \leftrightarrow \mathrm{TRI}(x, a, b)\,]$$

where $\mathrm{TRI}(x, y_0, y_1)$ is the condition

$$\Big(\forall S \in \Sigma\Big)\Big(\,(\{y_0, y_1\} \subseteq S \rightarrow x \in S) \;\wedge\; \bigwedge_{i=0,1} (x \in S \wedge y_i \notin S \longrightarrow$$

$$(\exists T \in \Sigma)\big[\,(S \subseteq T) \;\wedge\; (T \cap \{y_0, y_1\} = \{y_{1-i}\})\big]\Big)\Big).$$

PROOF. To prove this lemma, note first that $c \in (a/\zeta) \cup (b/\zeta)$ implies that $\mathrm{TRI}(c, a, b)$ holds. Thus it will suffice to find $\langle a, b \rangle \in U^2 - \zeta$ such that for every $c \in U$, $\mathrm{TRI}(c, a, b)$ implies $c \in (a/\zeta) \cup (b/\zeta)$.

We begin by choosing an arbitrary pair $\langle a, b \rangle \in U^2 - \zeta$. Assuming that this pair does not have the desired property, we then choose an element $c_2 \in U$ such that $\mathrm{TRI}(c_2, a, b)$ and $|\{a, b, c_2\}/\zeta| = 3$. We take $c_0 = a$ and $c_1 = b$. If the pair $\langle c_0, c_2 \rangle$ does not have the desired property, then we choose $c_3 \in U$ such that $\mathrm{TRI}(c_3, c_0, c_2)$ and $|\{c_0, c_2, c_3\}/\zeta| = 3$. Continuing in this way, we either find a pair of elements as desired, or we construct an infinite sequence

$$c_0, c_1, \ldots, c_n, \ldots \;(n < \omega)$$

with these properties:

$$\mathrm{TRI}(c_{i+1}, c_0, c_i) \text{ and } |\{c_0, c_i, c_{i+1}\}/\zeta| = 3 \text{ for all } 0 < i < \omega.$$

But it is not possible for such a sequence to exist, since the conditions imply that all the elements $c_0, c_1, \ldots, c_n, \ldots$ are distinct. That they are distinct follows easily from the assertion below, whose proof is left to the reader.

Claim. Suppose that $\mathrm{TRI}(w, u, v)$ and $\mathrm{TRI}(z, u, w)$, and that

$$|\{u, v, w\}/\zeta| = |\{u, w, z\}/\zeta| = 3.$$

Then $\mathrm{TRI}(z,u,v)$ and $|\{u,v,z\}/\zeta| = 3$.

The proof of the claim is fairly straightforward. (The proof that $\langle v,z\rangle \notin \zeta$ chiefly utilizes the assumption that $\langle v,w\rangle \notin \zeta$.) □

Proof of Lemma 2.4. Let $\mathbf{A}$ be a finite algebra and let U be a $Z_2(\mathbf{A})$-equivalence class that contains at least two $Z_1(\mathbf{A})$-equivalence classes. We shall write, throughout this proof, Z for $Z(\mathbf{A})$ ($= Z_1(\mathbf{A})$). In Lemma 2.11, we take for Σ the set Σ_1 of subsets of A defined in the proof of Lemma 2.3. As in the proof of Lemma 2.3, we choose $m+1$ -ary terms $t_1,\ldots,t_k$ such that

$$\Sigma = \left\{E^{\mathbf{A}}(t_i,\bar{c},\bar{d}) : 1 \le i \le k \ \text{ and } \ \bar{c},\bar{d} \in A^m\right\}. \tag{2.8}$$

Now the equivalence relation ζ on U defined in Lemma 2.11 is just Z restricted to U. Applying Lemma 2.11, we obtain a pair $\langle a,b\rangle$ of elements of U satisfying

$$\langle a,b\rangle \in U^2 - Z, \text{ and where } Q = (a/Z) \cup (b/Z)$$

$$\text{we have } Q = \{x \in U \ : \ \mathrm{TRI}(x,a,b)\}\ . \tag{2.9}$$

Our goal is to show that in every ae-closed diagonal subalgebra $\mathbf{D} \subseteq \mathbf{A}^X$, the set $D(Q)$ is definable. Let $\mathbf{D} \subseteq \mathbf{A}^X$ be such an algebra. It follows from Lemma 2.3 that the set $D(U)$ is definable in $\mathbf{D}$; in fact, $f \in D$ belongs to U^X iff $\mathbf{D} \models \mathrm{CEN}_2(\hat{a},f)$. The formula we construct to define $D(Q)$ in $\mathbf{D}$ will be a translation of "TRI", using $\hat{a}$ and $\hat{b}$ as parameters. We first note that, since A itself is among the members of Σ, $\mathrm{TRI}(x,y_0,y_1)$ is equivalent to the following formula, which is written in a form slightly more useful for our purposes.

$$\begin{aligned}\Big(\forall S \in \Sigma\Big)\Big(\big(\{y_0,y_1\} \subseteq S \to x \in S\big) \ \wedge\ \textstyle\bigwedge_{i=0,1}\Big(\exists T \in \Sigma\Big)\Big(\big(S \subseteq T\big) \ \wedge \\ \big(y_{1-i} \in T\big) \wedge \big((x \in S \wedge y_i \in T) \to y_i \in S\big)\Big)\Big).\end{aligned}$$

The formula $\mathrm{N}(x)$ written below incorporates the above formula, and will be seen to define $D(Q)$ in $\mathbf{D}$.

$$\mathrm{CEN}_2(\hat{a},x) \ \wedge\ \bigwedge_{1\le i\le k}(\forall\bar{u},\bar{v})\Big\{\big(\{\hat{a},\hat{b}\}\subseteq E(t_i,\bar{u},\bar{v})\rightarrow x\in E(t_i,\bar{u},\bar{v})\big)\ \&$$

$$\bigvee_{1\le j\le k}(\exists\bar{u}',\bar{v}')\Big(\big(E(t_i,\bar{u},\bar{v})\subseteq E(t_j,\bar{u}',\bar{v}')\big)\ \wedge\ \big(\hat{a}\in E(t_j,\bar{u}',\bar{v}')\big)\wedge$$
$$\big((x\in E(t_i,\bar{u},\bar{v})\ \wedge\ \hat{b}\in E(t_j,\bar{u}',\bar{v}'))\rightarrow\hat{b}\in E(t_i,\bar{u},\bar{v})\big)\Big)\quad\&$$

$$\bigvee_{1\le j\le k}(\exists\bar{u}',\bar{v}')\Big(\big(E(t_i,\bar{u},\bar{v})\subseteq E(t_j,\bar{u}',\bar{v}')\big)\ \wedge\ \big(\hat{b}\in E(t_j,\bar{u}',\bar{v}')\big)\wedge$$
$$\big((x\in E(t_i,\bar{u},\bar{v})\ \wedge\ \hat{a}\in E(t_j,\bar{u}',\bar{v}'))\rightarrow\hat{a}\in E(t_i,\bar{u},\bar{v})\big)\Big)\Big\}$$

To see that this formula works, suppose first that $\alpha\in D(Q)$. Thus $\alpha\in D(U)$, and so $\mathbf{D}\models\mathrm{CEN}_2(\hat{a},\alpha)$. To check the validity of the second clause of the formula $\mathrm{N}(x)$ for $x=\alpha$, let $\bar{f},\bar{g}\in D^m$ be such that $\hat{a},\hat{b}\in E^{\mathbf{D}}(t_i,\bar{f},\bar{g})$. This means that for every $x\in X$ and $u\in\{a,b\}$,

$$t_i^{\mathbf{A}}(u,\bar{f}(x))=t_i^{\mathbf{A}}(u,\bar{g}(x)).$$

Since $\alpha\in D(Q)$, we have $\alpha(x)Za$ or $\alpha(x)Zb$, for every $x\in X$. Then it follows from the above equation, and from the definition of Z, that

$$t_i^{\mathbf{A}}(\alpha(x),\bar{f}(x))=t_i^{\mathbf{A}}(\alpha(x),\bar{g}(x)),$$

and this holds for every x; i.e., $\alpha\in E^{\mathbf{D}}(t_i,\bar{f},\bar{g})$.

Continuing with the demonstration that $\mathbf{D}\models\mathrm{N}(\alpha)$, we have to check the third and fourth clauses of $\mathrm{N}(x)$. They are both similar, so we just consider the third clause. Suppose first that either $\hat{b}\in E^{\mathbf{D}}(t_i,\bar{f},\bar{g})$ or $\alpha\notin E^{\mathbf{D}}(t_i,\bar{f},\bar{g})$. In this case, we can satisfy the condition in a trivial way by choosing $j=i$ and $\bar{f}'=\bar{g}'=\bar{f}$. Now suppose that $\hat{b}\notin E^{\mathbf{D}}(t_i,\bar{f},\bar{g})$ and $\alpha\in E^{\mathbf{D}}(t_i,\bar{f},\bar{g})$. Then we can choose x_0 such that

$$t_i^{\mathbf{A}}(b,\bar{f}(x_0))\neq t_i^{\mathbf{A}}(b,\bar{g}(x_0)),\quad\text{while}\quad t_i^{\mathbf{A}}(\alpha(x_0),\bar{f}(x_0))=t_i^{\mathbf{A}}(\alpha(x_0),\bar{g}(x_0)).$$

Since $\alpha(x_0)\in Q$ (and by (2.8) and (2.9)), we can now choose $1\le j\le k$ and $\bar{c},\bar{d}\in A^m$ so that $b\notin E^{\mathbf{A}}(t_j,\bar{c},\bar{d})$, $a\in E^{\mathbf{A}}(t_j,\bar{c},\bar{d})$, and

$$E^{\mathbf{A}}(t_i,\bar{f}(x_0),\bar{g}(x_0))\subseteq E^{\mathbf{A}}(t_j,\bar{c},\bar{d}).$$

We define $\bar{f}'$ so that $\bar{f}'(x)=\bar{c}$ for all x, and define $\bar{g}'$ so that $\bar{g}'(x)=\bar{c}$ for $x\neq x_0$ and $\bar{g}'(x_0)=\bar{d}$. Since D is ae-closed, these functions belong to D. A brief check of our definitions should convince the reader that where $S=E^{\mathbf{D}}(t_j,\bar{f}',\bar{g}')$, we have $E^{\mathbf{D}}(t_i,\bar{f},\bar{g})\subseteq S$ and $\hat{b}\notin S$, while $\hat{a}\in S$; and so we have satisfied the third clause of $\mathrm{N}(x)$ for $x=\alpha$.

We have completed the verification that $\alpha \in D(Q)$ implies $\mathbf{D} \models \mathrm{N}(\alpha)$. To conclude this proof, we must show that the converse holds. To do this, let us suppose that $\alpha \in D - D(Q)$. We assume also that $\alpha \in D(U)$, since otherwise it's clear that $\mathbf{D} \models \neg\mathrm{N}(\alpha)$. Choose x_0 so that $\alpha(x_0) \notin Q$. By (2.9), we have $\neg\mathrm{TRI}(\alpha(x_0), a, b)$. One of the clauses of the condition "TRI" fails for $x = \alpha(x_0)$, $y_0 = a$, $y_1 = b$. We need to show that some clause of $\mathrm{N}(x)$ fails for $x = \alpha$. All the arguments are rather similar; so we shall only deal explicitly with this case:

> Where $u = \alpha(x_0)$, we have elements $\bar{c}, \bar{d}$ such that
> $u \in E^{\mathbf{A}}(t_i, \bar{c}, \bar{d})$ and $a \notin E^{\mathbf{A}}(t_i, \bar{c}, \bar{d})$.
> There do not exist j and $\bar{c}', \bar{d}'$ such that
> $E^{\mathbf{A}}(t_i, \bar{c}, \bar{d}) \subseteq E^{\mathbf{A}}(t_j, \bar{c}', \bar{d}')$ and $\{a, b\} \cap E^{\mathbf{A}}(t_j, \bar{c}', \bar{d}') = \{b\}$.

Assuming the above, define $\bar{f}, \bar{g} \in D^m$ so that $\bar{f}(x) = \bar{c}$ for all x, and $\bar{g}(x) = \bar{c}$ for $x \neq x_0$, while $\bar{g}(x_0) = \bar{d}$. Now it is clear that $\alpha \in E^{\mathbf{D}}(t_i, \bar{f}, \bar{g})$ and $\hat{a} \notin E^{\mathbf{D}}(t_i, \bar{f}, \bar{g})$. It is also easy to see that we have an explicit failure of the last clause of the formula $\mathrm{N}(\alpha)$. Namely, suppose that $\bar{f}', \bar{g}' \in D^m$ are such that $E^{\mathbf{D}}(t_i, \bar{f}, \bar{g}) \subseteq E^{\mathbf{D}}(t_j, \bar{f}', \bar{g}')$. Now if $\beta \in D$ then $\beta \in E^{\mathbf{D}}(t_i, \bar{f}, \bar{g})$ iff $\beta(x_0) \in E^{\mathbf{D}}(t_i, \bar{c}, \bar{d})$. Since $\mathbf{D}$ is ae-closed, an easy argument shows that $\beta \in E^{\mathbf{D}}(t_j, \bar{f}', \bar{g}')$ iff $\beta(x_0) \in E^{\mathbf{D}}(t_j, \bar{f}'(x_0), \bar{g}'(x_0))$. Therefore, $E^{\mathbf{A}}(t_i, \bar{c}, \bar{d}) \subseteq E^{\mathbf{A}}(t_j, \bar{f}'(x_0), \bar{g}'(x_0))$. The assumptions displayed above then imply that if $\hat{b} \in E^{\mathbf{D}}(t_j, \bar{f}', \bar{g}')$ then also $\hat{a} \in E^{\mathbf{D}}(t_j, \bar{f}', \bar{g}')$. Hence $E^{\mathbf{D}}(t_j, \bar{f}', \bar{g}')$ cannot fulfill the last clause of $\mathrm{N}(x)$ for $x = \alpha$. With these remarks, we have concluded our proof of Lemma 2.4. □

Chapter 3

Centerless algebras

We continue to work with a fixed locally finite variety $\mathcal{V}$ that is structured. This chapter is devoted to a proof of the theorem below.

THEOREM 3.1 *Let* $\mathbf{F}$ *be a finite, subdirectly irreducible, centerless algebra in* $\mathcal{V}$. *Every subalgebra of* $\mathbf{F}$ *having at least two elements is simple and non-Abelian.*

Before launching into the proof, we pause to establish a useful corollary of Theorems 2.1 and 3.1.

COROLLARY 3.2 *If* $\mathbf{A} \in \mathcal{V}$ *then* $Z(\mathbf{A})$ *is the largest locally solvable congruence of* $\mathbf{A}$. *Every locally solvable algebra in* $\mathcal{V}$ *is Abelian.*

PROOF. Suppose that $\mathbf{A} \in \mathcal{V}$ and that δ is a locally solvable congruence of $\mathbf{A}$. For the sake of deriving a contradiction, let us suppose that $\delta \not\subseteq Z(\mathbf{A})$. Letting $\langle u, v\rangle \in \delta - Z(\mathbf{A})$, we can choose a finitely generated algebra $\mathbf{B} \subseteq \mathbf{A}$ with $\langle u, v\rangle \in B^2 - Z(\mathbf{B})$. Now $\mathbf{B}$ is finite since $\mathcal{V}$ is locally finite, and the congruence $\delta|_B$ is a solvable congruence of $\mathbf{B}$ not contained in $Z(\mathbf{B})$. Using Theorem 0.11, we derive that $\delta|_B \vee Z(\mathbf{B})$ is solvable over $Z(\mathbf{B})$; and so we can choose a congruence μ such that $Z(\mathbf{B}) \prec \mu$ and μ is Abelian over $Z(\mathbf{B})$. Let η be any congruence of $\mathbf{B}$ maximal in the set of congruences χ such that $Z(\mathbf{B}) \subseteq \chi$ and $\mu \not\subseteq \chi$. Then $\eta \wedge \mu = Z(\mathbf{B})$; and letting $\hat{\eta} = \eta \vee \mu$, we have that $\eta \prec \hat{\eta}$.

Now the algebra $\mathbf{F} = \mathbf{B}/\eta$ is a finite subdirectly irreducible algebra in $\mathcal{V}$ with monolith $\beta = \hat{\eta}/\eta$. We have $\operatorname{typ}(\eta, \hat{\eta}) = \operatorname{typ}(Z(\mathbf{B}), \mu)$ by Lemma 0.37; hence $\hat{\eta}$ is Abelian over η, which implies that the monolith β is an Abelian congruence of $\mathbf{F}$. By Theorem 2.1, the algebra $\mathbf{B}/Z(\mathbf{B})$ is centerless, implying that μ does not centralize 1_B over $Z(\mathbf{B})$; and this implies (in the situation of Figure 3.1) that $\hat{\eta}$ does not centralize 1_B over η.

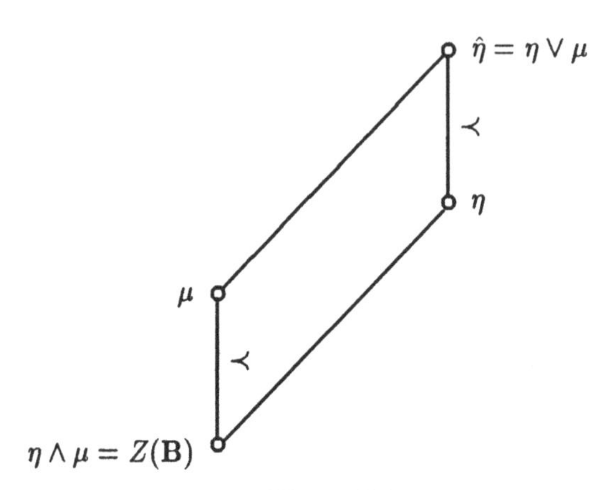

Figure 3.1

Thus the monolith of $\mathbf{F}$ is not contained in its center, which implies that $Z(\mathbf{F}) = 0_F$. Thus $\mathbf{F}$ is a finite, subdirectly irreducible, centerless algebra. By Theorem 3.1, $\mathbf{F}$ is simple, implying that $\hat{\eta} = 1_B$. But then $\hat{\eta}$ does centralize 1_B over η, because $\hat{\eta}$ is Abelian over η. This contradiction shows that every locally solvable congruence of $\mathbf{A}$ is contained in $Z(\mathbf{A})$. Of course, $Z(\mathbf{A})$ is locally solvable (even Abelian).

If $\mathbf{A} \in \mathcal{V}$ is locally solvable then 1_A is a locally solvable congruence, so $1_A = Z(\mathbf{A})$ by what we just proved, and thus $\mathbf{A}$ is Abelian. □

Our proof of Theorem 3.1 is broken into three lemmas, 3.3, 3.4, and 3.5. In the next chapter we shall need to consider subdirect products in which the factors are not isomorphic. Therefore we state the first lemma in a more general form than is required for the proof of Theorem 3.1.

Suppose that $\mathbf{D} \leq \prod_{x \in X} \mathbf{B}_x$. We extend to $\mathbf{D}$ the notation $[\![f = g]\!]$ and related notations and concepts introduced in Definition 0.43 (whenever they make sense). For example, we say that $\mathbf{D}$ is **ae-closed** iff for every $f \in D$ and $g \in \prod_{x \in X} B_x$ such that $[\![f \neq g]\!]$ is a finite set, we have $g \in D$.

LEMMA 3.3 *Let $\mathbf{B}_1, \ldots, \mathbf{B}_n$ be finite, subdirectly irreducible, centerless algebras of the same type. There exist formulas $\mathrm{P}_{\subseteq}(x, y, u, v)$ and $\min(x, y)$ satisfying: whenever $\mathbf{D} \leq \prod_{x \in X} \mathbf{B}_x$ is an ae-closed subdirect product with $\{\mathbf{B}_x : x \in X\} \subseteq \{\mathbf{B}_1, \ldots, \mathbf{B}_n\}$, and $\{f, g, h, k\} \subseteq D$, then*

$$\mathbf{D} \models \mathrm{P}_{\subseteq}(f, g, h, k) \text{ iff } [\![f = g]\!] \subseteq [\![h = k]\!], \text{ and}$$

$$\mathbf{D} \models \min(f, g) \text{ iff } |[\![f \neq g]\!]| = 1.$$

PROOF. Let $\mathbf{D} \leq \prod_{x \in X} \mathbf{B}_x$ be an ae-closed subdirect product with $\{\mathbf{B}_x : x \in X\} \subseteq \{\mathbf{B}_1, \ldots, \mathbf{B}_n\}$. The formulas we construct will not depend in any way on this choice of $\mathbf{D}$. We begin by choosing $q+1$-ary terms (for some q) $t_0, \ldots, t_{k-1}$ such that whenever $\{a, b\} \subseteq B \in \{B_1, \ldots, B_n\}$ and $a \neq b$, there exist $i < k$ and $\bar{c}, \bar{d} \in B^q$ such that

$$t_i^{\mathbf{B}}(a, \bar{c}) = t_i^{\mathbf{B}}(a, \bar{d}) \leftrightarrow t_i^{\mathbf{B}}(b, \bar{c}) \neq t_i^{\mathbf{B}}(b, \bar{d}).$$

(We can do this since the $\mathbf{B}_i$ are finite and centerless.) We shall write $x \in E_i(\bar{y}, \bar{z})$ as shorthand for the formula $t_i(x, \bar{y}) \approx t_i(x, \bar{z})$. Now for each $i < k$ we define a formula of $2q+2$ variables.

$$\Gamma_i(x, y, \bar{u}, \bar{v}) \overset{\text{def}}{\longleftrightarrow} x \in E_i(\bar{u}, \bar{v}) \ \wedge \ y \notin E_i(\bar{u}, \bar{v}) \wedge$$

$$(\forall \bar{u}', \bar{v}')[(E_i(\bar{u}, \bar{v}) \subseteq E_i(\bar{u}', \bar{v}') \ \wedge \ y \notin E_i(\bar{u}', \bar{v}')) \rightarrow E_i(\bar{u}, \bar{v}) = E_i(\bar{u}', \bar{v}')].$$

These formulas have the following properties. Let $\mathbf{D}$ be as above and suppose that $f, g \in D$.

For all $x_0 \in X, f(x_0) \neq g(x_0)$ iff for some $i < k$ there exist $\bar{h}, \bar{k} \in D^q$ such that $\bar{h}(x) = \bar{k}(x)$ for all $x \neq x_0$ and $\mathbf{D} \models \Gamma_i(f, g, \bar{h}, \bar{k}) \vee \Gamma_i(g, f, \bar{h}, \bar{k})$. (3.1)

$\mathbf{D} \models \Gamma_i(f, g, \bar{h}, \bar{k})$ iff there exists $x_0 \in X$ such that: $\mathbf{B}_{x_0} \models \Gamma_i(f(x_0), g(x_0), \bar{h}(x_0), \bar{k}(x_0))$ (implying that $f(x_0) \neq g(x_0)$); and for all x, $E_i^{\mathbf{B}_x}(\bar{h}(x), \bar{k}(x)) \neq B_x$ iff $x = x_0$. For this x_0, and for any $\alpha \in D$, we have that $\alpha \in E_i^{\mathbf{D}}(\bar{h}, \bar{k})$ iff $\alpha(x_0) \in E_i^{\mathbf{B}_{x_0}}(\bar{h}(x_0), \bar{k}(x_0))$. (3.2)

We sketch a proof of (3.1) and leave (3.2) for the reader to prove. Let $f(x_0) \neq g(x_0)$. Write $\mathbf{B}$ for $\mathbf{B}_{x_0}$. Choose an i and $\bar{c}, \bar{d} \in B^q$ such that

$$f(x_0) \in E_i^{\mathbf{B}}(\bar{c}, \bar{d}) \leftrightarrow g(x_0) \notin E_i^{\mathbf{B}}(\bar{c}, \bar{d}).$$

We can assume that the set $E_i^{\mathbf{B}}(\bar{c}, \bar{d})$ is maximal with this property; i.e., if $\bar{u}, \bar{v} \in B^q$ and $E_i^{\mathbf{B}}(\bar{c}, \bar{d})$ is properly contained in $E_i^{\mathbf{B}}(\bar{u}, \bar{v})$, then the latter set contains both $f(x_0)$ and $g(x_0)$. Without losing generality, we can assume further that $f(x_0) \in E_i^{\mathbf{B}}(\bar{c}, \bar{d})$ and $g(x_0) \notin E_i^{\mathbf{B}}(\bar{c}, \bar{d})$. Now it follows that $\mathbf{B} \models \Gamma_i(f(x_0), g(x_0), \bar{c}, \bar{d})$. Since D is ae-closed, we can choose $\bar{h}, \bar{k} \in D^q$ so that $[\![\bar{h} = \bar{k}]\!] = X - \{x_0\}$ and $\bar{h}(x_0) = \bar{c}$, $\bar{k}(x_0) = \bar{d}$. It is easy to see that $\mathbf{D} \models \Gamma_i(f, g, \bar{h}, \bar{k})$. The converse assertion in (3.1) is easier.

Up to this point, we have not used the hypothesis that the $\mathbf{B}_i$ are subdirectly irreducible. We shall now do so. Let m be a finite upper bound to the cardinalities of all the algebras $\mathbf{B}_1, \ldots, \mathbf{B}_n$. Then every binary polynomial operation of an algebra $\mathbf{B}_i$ takes the form $H(x, y) = p^{\mathbf{B}_i}(x, y, \bar{b})$ for some $m + 2$ -ary term p and some $\bar{b} \in B_i^m$. Since each of these (finite) algebras has only finitely many binary polynomial operations, we can find finitely many $m + 2$ -ary terms $p_0(x, y, \bar{z}), \ldots, p_{r-1}(x, y, \bar{z})$ such that every binary term operation of any algebra $\mathbf{B}_i$ takes the form $p_j^{\mathbf{B}_i}(x, y, \bar{b})$ for some $j < r$ and some $\bar{b} \in B_i^m$.

Now let $\mathbf{B}$ be any one of the algebras $\mathbf{B}_0, \ldots, \mathbf{B}_{n-1}$. Choose a pair $\langle a, b\rangle$ in the monolith of $\mathbf{B}$ with $a \neq b$. The fact that $\langle a, b\rangle$ belongs to the monolith means that for all $x, y \in B$, if $x \neq y$ then $\langle a, b\rangle \in \mathrm{Cg}^{\mathbf{B}}(x, y)$. This is equivalent to the following assertion.

For all $x \neq y$ in B there exist binary polynomial operations $P_0, \ldots, P_s$ such that $a = P_0(x, y)$, $b = P_s(y, x)$, and for all $j < s$, $P_j(y, x) = P_{j+1}(x, y)$.

(Actually, we can choose each P_j to be independent either of its first, or its second, variable.) Now applying (3.1) to the algebra $\mathbf{B}$ itself, we find an $i < k$, and $\bar{c}, \bar{d} \in B^q$ such that $\mathbf{B} \models \Gamma_i(a, b, \bar{c}, \bar{d}) \vee \Gamma_i(b, a, \bar{c}, \bar{d})$. Thus in particular we have that

$$a \in E_i^{\mathbf{B}}(\bar{c}, \bar{d}) \leftrightarrow b \notin E_i^{\mathbf{B}}(\bar{c}, \bar{d}).$$

From the two assertions displayed above, it follows that for any two unequal elements x and y in B, there exists a binary polynomial operation P, or equivalently some $j < r$ and $\bar{b} \in B^m$, so that

$$p_j^{\mathbf{B}}(x, y, \bar{b}) \in E_i^{\mathbf{B}}(\bar{c}, \bar{d}) \leftrightarrow p_j^{\mathbf{B}}(y, x, \bar{b}) \notin E_i^{\mathbf{B}}(\bar{c}, \bar{d}).$$

Therefore, for every $l \in \{1, \ldots, n\}$ the formula

$$\bigvee_{i<k} (\exists a, b, \bar{c}, \bar{d})\Big[\Gamma_i(a, b, \bar{c}, \bar{d}) \wedge (\forall x, y)\Big\{\neg(x \approx y) \longleftrightarrow \tag{3.3}$$

$$\bigvee_{j<r} (\exists \bar{b})(p_j(x, y, \bar{b}) \in E_i(\bar{c}, \bar{d}) \leftrightarrow p_j(y, x, \bar{b}) \notin E_i(\bar{c}, \bar{d}))\Big\}\Big]$$

is valid in $\mathbf{B}_l$.

We now define two more auxiliary formulas.

$$\Pi(x,y,u,v) \stackrel{\text{def}}{\longleftrightarrow} \bigvee_{j<r}(\exists\bar{y})(x \approx p_j(u,v,\bar{y}) \wedge y \approx p_j(v,u,\bar{y}))$$

$$\begin{aligned} \mathrm{N}(x,y,u,v) \stackrel{\text{def}}{\longleftrightarrow} \bigvee_{i<k}(\exists x',y',u',v',w,z,\bar{w},\bar{z}) \\ \Big[\Pi(x',y',x,y) \wedge \Pi(u',v',u,v) \wedge \Gamma_i(w,z,\bar{w},\bar{z}) \wedge \\ (x' \in E_i(\bar{w},\bar{z}) \leftrightarrow y' \notin E_i(\bar{w},\bar{z})) \wedge \\ (u' \in E_i(\bar{w},\bar{z}) \leftrightarrow v' \notin E_i(\bar{w},\bar{z}))\Big]. \end{aligned}$$

The second formula has the following property, for all $f,g,h,k \in D$.

$$\mathbf{D} \models \mathrm{N}(f,g,h,k) \text{ iff } [\![f \neq g]\!] \cap [\![h \neq k]\!] \neq \emptyset. \tag{3.4}$$

To prove (3.4), we assume first that $\mathbf{D} \models \mathrm{N}(f,g,h,k)$. Let f', g', h', k', w, $z \in D$, $i < k$, and $\bar{w},\bar{z} \in D^q$ be the elements whose existence is asserted in the formula. Notice that since $\mathbf{D} \models \Pi(f',g',f,g)$, we have $[\![f = g]\!] \subseteq [\![f' = g']\!]$. Similarly, $[\![h = k]\!] \subseteq [\![h' = k']\!]$. Thus it will be enough to show that $[\![f' \neq g']\!] \cap [\![h' \neq k']\!] \neq \emptyset$. Since $\mathbf{D} \models \Gamma_i(w,z,\bar{w},\bar{z})$, then by (3.2) there exists $x_0 \in X$ such that for any $\alpha \in D$ we have $\alpha \in E_i^{\mathbf{D}}(\bar{w},\bar{z})$ iff $\alpha(x_0) \in E_i^{\mathbf{B}_{x_0}}(\bar{w}(x_0),\bar{z}(x_0))$. Now the last two clauses of the formula $\mathrm{N}(f,g,h,k)$ imply that $x_0 \in [\![f' \neq g']\!] \cap [\![h' \neq k']\!]$.

To prove the converse assertion in (3.4), we assume that $f,g,h,k \in D$ and $x_0 \in [\![f \neq g]\!] \cap [\![h \neq k]\!]$. Since the formula (3.3) holds in $\mathbf{B}_{x_0}$ we choose $i < k$ and $a,b \in B_{x_0}$ and $\bar{c},\bar{d} \in B_{x_0}^q$ to satisfy it. Then we choose $\bar{w}$, $\bar{z} \in D^q$ so that $[\![\bar{w} = \bar{z}]\!] = X - \{x_0\}$ and $\bar{w}(x_0) = \bar{c}$ and $\bar{z}(x_0) = \bar{d}$. And we choose $w,z \in D$ such that $w(x_0) = a$ and $z(x_0) = b$. Elements with these properties exist because $\mathbf{D}$ is ae-closed. Now $\mathbf{D} \models \Gamma_i(w,z,\bar{w},\bar{z})$ by (3.2), because $\mathbf{B}_{x_0} \models \Gamma_i(a,b,\bar{c},\bar{d})$. It remains to find $f',\dots,k'$ to verify that $\mathbf{D} \models \mathrm{N}(f,g,h,k)$.

By (3.3) there exist $j < r$ and $\bar{y} \in D^m$ such that

$$\begin{aligned} p_j^{\mathbf{B}_{x_0}}(f(x_0),g(x_0),\bar{y}(x_0)) \in E_i^{\mathbf{B}_{x_0}}(\bar{w}(x_0),\bar{z}(x_0)) \longleftrightarrow \\ p_j^{\mathbf{B}_{x_0}}(g(x_0),f(x_0),\bar{y}(x_0)) \notin E_i^{\mathbf{B}_{x_0}}(\bar{w}(x_0),\bar{z}(x_0)). \end{aligned}$$

Since $\bar{w}(x) = \bar{z}(x)$ for $x \neq x_0$, this means that where

$$f' = p_j^{\mathbf{D}}(f,g,\bar{y}) \text{ and } g' = p_j^{\mathbf{D}}(g,f,\bar{y})$$

we have

$$f' \in E_i^{\mathbf{D}}(\bar{w},\bar{z}) \text{ iff } g' \notin E_i^{\mathbf{D}}(\bar{w},\bar{z}).$$

In the same fashion, we can find h' and k' corresponding to h and k. It should be clear that $\mathbf{D} \models \mathrm{N}(f,g,h,k)$ by taking f',g',h',k' for x',y',u',v' respectively in the formula.

Finally, we are in position to define the desired formulas and complete the proof of this lemma. We take

$$\mathrm{P}_{\subseteq}(x,y,u,v) \overset{\text{def}}{\longleftrightarrow} (\forall z,w)(\mathrm{N}(z,w,u,v) \rightarrow \mathrm{N}(z,w,x,y)).$$

$$\min(u,v) \overset{\text{def}}{\longleftrightarrow} \neg(u \approx v) \wedge$$

$$(\forall x,y)[\{\neg(x \approx y) \wedge \mathrm{P}_{\subseteq}(u,v,x,y)\} \rightarrow \mathrm{P}_{\subseteq}(x,y,u,v)].$$

Using (3.4), it is easy to verify that these formulas do the job. □

LEMMA 3.4 *Let* $\mathbf{F}$ *be a finite, subdirectly irreducible, centerless algebra. Suppose that* $\mathbf{F}$ *has a subalgebra which possesses three or more congruences. Then the class* $\boldsymbol{P}_{\mathbf{s}}(\mathbf{F})$ *is unstructured.*

PROOF. This is immediate from Theorem 0.48 and Lemma 3.3. □

LEMMA 3.5 *Let* $\mathbf{F}$ *be a finite, subdirectly irreducible, centerless algebra. Suppose that* $\mathbf{F}$ *has an Abelian subalgebra which possesses more than one element. Then the class* $\boldsymbol{P}_{\mathbf{s}}(\mathbf{F})$ *is unstructured.*

PROOF. Let $\mathbf{A}$ be an Abelian subalgebra of $\mathbf{F}$ containing elements $0,1$ where $0 \neq 1$. We assume that $\mathbf{F}$ is finite, subdirectly irreducible, and centerless.

The proof of this lemma is nearly the same as the proof of Theorem 2.1, and is simpler in some respects. It amounts to a reworking of the proofs of Lemmas 2.6 and 2.9, including minor changes in the definitions of the algebras used in the interpretations, and different proofs for some of the details.

We put $Q = \{0,1\}$. Then the proof divides naturally into two cases, depending on whether the two-element minimal algebra $\mathbf{R} = \mathbf{A}|_Q$ is of type **1** or **2**. The class $\mathcal{C}_1$ (Definition 2.5 and Lemma 2.7) can be interpreted into $\boldsymbol{P}_{\mathbf{s}}(\mathbf{F})$ in the first case; the class $\mathcal{C}_2$ (Definition 2.8 and Lemma 2.10) in the second case. We shall give details here only for the first case, in which $\mathbf{A}|_Q$ is essentially unary. Some remarks on the second case will be found at the end of the proof.

Thus suppose $\mathbf{A}|_Q$ is essentially unary. Continuing to follow the proof of Lemma 2.6, we start with a structure $\mathbf{S} \in \mathcal{C}_1$, $S \subseteq \{0,1\}^X$. We define $T = X \cup Q^S$; and we put

$$X^* = \bigcup\{X_t : t \in T\} \text{ and } Y^* = \bigcup\{X_t : t \in X\},$$

where every set X_t is infinite and $X_t \cap X_{t'} = \emptyset$ when $t \neq t'$. We define the functions $s^* \in Q^{X^*}$ corresponding to $s \in S$, and the function $\mu \in Q^{X^*}$, just as before (by formulas (2.1) and (2.2)). Let $\mathbf{D}'$ be the subalgebra of $\mathbf{A}^{X^*}$ generated by the set consisting of all the functions s^* for $s \in S$, of μ, and of all the constant functions $\hat{a}$ for $a \in A$. Then let $\mathbf{D}$ be the ae-closure of $\mathbf{D}'$ in $\mathbf{A}^{X^*}$; and let $\mathbf{H}$ be the ae-closure of $\mathbf{D}'$ in $\mathbf{F}^{X^*}$. Thus $\mathbf{D}$ is an ae-closed diagonal subalgebra of $\mathbf{A}^{X^*}$, while $\mathbf{H}$ is an ae-closed, but not a diagonal, subalgebra of $\mathbf{F}^{X^*}$.

In the proof of Lemma 2.6, we observed that $\mathbf{S}$ is isomorphic to the structure $\mathbf{L}' = \langle L', \leq, \text{MEET}, \text{JOIN}\rangle$ in $\mathcal{C}$ induced by the inclusion $L' \subseteq \{0,1\}^{Y^*}$, where $L' = \{s^*|_{Y^*} : s \in S\}$. We saw that where L is the ae-closure of L' in $\{0,1\}^{Y^*}$, the structure $\mathbf{L}$ in $\mathcal{C}$ induced by the inclusion $L \subseteq \{0,1\}^{Y^*}$ interprets $\mathbf{S}$. Thus all we need to do in this proof is to show that $\mathbf{L}$ can be interpreted in $\mathbf{H}$, using the parameters $\hat{a}$ ($a \in A$). Since $\mathbf{H}$ is an ae-closed subalgebra of $\mathbf{F}^{X^*}$, we can use the formulas $\mathrm{P}_{\subseteq}(x,y,u,v)$ and $\min(x,y)$ that were introduced in Lemma 3.3.

Suppose that $A = \{a_0, \ldots, a_{k-1}\}$ with $a_0 = 0$ and $a_1 = 1$, and put

$$\bar{a}' = \langle \hat{a}_0, \ldots, \hat{a}_{k-1}\rangle \in H^k .$$

Note that $D = H(A)$ ($= H \cap A^{X^*}$) and $D(Q) = H(Q)$. Here are formulas that define D and $D(Q)$ in $\mathbf{H}$.

$$\mathrm{D}(x, y_0, \ldots, y_{k-1}) \overset{\text{def}}{\longleftrightarrow} \neg(\exists u, v)\big[\min(u,v) \wedge \bigwedge_{i<k} \mathrm{P}_{\subseteq}(x, y_i, u, v)\big].$$

$$\mathrm{Q}(x, y_0, \ldots, y_{k-1}) \overset{\text{def}}{\longleftrightarrow} \neg(\exists u, v)\big[\min(u,v) \wedge \bigwedge_{i<2} \mathrm{P}_{\subseteq}(x, y_i, u, v)\big].$$

For $f \in H$, we have $f \in D$ iff $\mathbf{H} \models \mathrm{D}(f, \bar{a}')$ and $f \in D(Q)$ iff $\mathbf{H} \models \mathrm{Q}(f, \bar{a}')$; as follows immediately from Lemma 3.3. The next formula defines the Boolean order on $D(Q)$.

$$\mathrm{LE}(u, v, \bar{y}) \overset{\text{def}}{\longleftrightarrow} \mathrm{Q}(u, \bar{y}) \wedge \mathrm{Q}(v, \bar{y}) \wedge \mathrm{P}_{\subseteq}(u, y_1, v, y_1).$$

For $f, g \in D(Q)$ we have $[\![f \leq g]\!] = X^*$ iff $\mathbf{H} \models \mathrm{LE}(f, g, \bar{a}')$. The proof of this assertion is also immediate from Lemma 3.3.

Following the proof of Lemma 2.6, we let $F : D(Q) \to \{0,1\}^{Y^*}$ be the restriction map, i.e., $F(f) = f|_{Y^*}$. Then the relation $F(f) \leq F(g)$ is defined in $\mathbf{H}$ by the formula

$$\mathrm{LE}^{\mu}(u, v, \bar{y}) \overset{\text{def}}{\longleftrightarrow} \mathrm{Q}(u, \bar{y}) \wedge \mathrm{Q}(v, \bar{y}) \wedge (\forall z, w)$$

$$\Big[\big\{\min(z,w) \wedge \mathrm{P}_{\subseteq}(u, y_0, z, w) \wedge \mathrm{P}_{\subseteq}(\mu, y_0, z, w)\big\} \to \mathrm{P}_{\subseteq}(v, y_0, z, w)\Big].$$

It is easily verified that $\mathbf{H} \models \mathrm{LE}^{\mu}(f, g, \bar{a}')$ iff $f, g \in D(Q)$ and $F(f) \leq F(g)$.

Of course, the kernel of F is also definable. It should be obvious, moreover, that the F-inverse images of the ternary relations MEET and JOIN in Q^{Y^*} are definable in $\mathbf{H}$ by formulas constructed in a similar way to the above. Thus the structure

$$\mathbf{K} = \langle K, \leq, \text{MEET}, \text{JOIN} \rangle$$

is interpretable in $\mathbf{H}$, where $K = F(D(Q))$. A proof that $K = L$, the ae-closure of $L' = \{s^*|_{Y^*} : s \in S\}$, can be given by following almost verbatim the proof of Claim 2 in Lemma 2.6. [In place of the observation that $f(0, \bar{y})$ $(\bar{y} \in Q^k)$ takes on only two values modulo Z_1, one observes that $f(0, \bar{y})$ $(\bar{y} \in Q^k)$ takes only values in $\{0, 1\}$. Since $\mathbf{A}$ is Abelian, it follows that $f(1, \bar{y})$ $(\bar{y} \in Q^k)$ takes on only two values. Since it does take the values 0 and 1, it follows that Q is closed under $f(1, \bar{y})$.]

This concludes our proof of Lemma 3.5 in the case where $\mathbf{A}|_Q$ is essentially unary. In the other case, where this algebra is polynomially equivalent with a two-element group, the proof is obtained by modifying the proof of Lemma 2.9 in precisely the same way that we modified the proof of Lemma 2.6 above. No new ideas are needed for this case. □

Theorem 3.1 is an immediate consequence of Lemmas 3.4 and 3.5.

Chapter 4

The discriminator subvariety

We continue to work with a fixed locally finite variety $\mathcal{V}$ that is structured. Recall from Definition 1.1 that $\mathcal{S}_3$ is the class of finite subdirectly irreducible algebras in $\mathcal{V}$ with type **3** monolith, and that $\mathcal{V}_3$ is the variety generated by $\mathcal{S}_3$. This chapter is devoted to a proof of the theorem below, which is one of the two principal results in Part I. A good portion of the proof of this theorem has already been given in Chapter 3.

THEOREM 4.1 *$\mathcal{V}_3$ is a discriminator variety.*

COROLLARY 4.2 *$\mathcal{V}_3$ is identical with the class of centerless algebras in $\mathcal{V}$.*

PROOF. Since $\mathcal{V}_3$ is a discriminator variety, every algebra in $\mathcal{V}_3$ is centerless. (See Theorem 0.22.) Now let $\mathbf{A} \in \mathcal{V}$ be centerless. For the sake of argument, assume that $\mathbf{A} \notin \mathcal{V}_3$. Then there exists an equation $p(\bar{x}) \approx q(\bar{x})$ valid in $\mathcal{V}_3$ which fails to hold in $\mathbf{A}$. Choose $\bar{a} = \langle a_0, \ldots, a_{n-1}\rangle$ such that $p^{\mathbf{A}}(\bar{a}) \neq q^{\mathbf{A}}(\bar{a})$. Since $\mathbf{A}$ is centerless, then where $e = p^{\mathbf{A}}(\bar{a})$ and $f = q^{\mathbf{A}}(\bar{a})$, we can choose a $k+1$ -ary term $t(x, \bar{y})$ (for some k) and $\bar{u}, \bar{v} \in A^k$ such that

$$t^{\mathbf{A}}(e, \bar{u}) = t^{\mathbf{A}}(e, \bar{v}) \leftrightarrow t^{\mathbf{A}}(f, \bar{u}) \neq t^{\mathbf{A}}(f, \bar{v}).$$

Let $\mathbf{B}$ be the subalgebra of $\mathbf{A}$ generated by $a_0, \ldots, a_{n-1}$ and the u_i and the v_i. Then $\mathbf{B}$ is finite and we have $\langle e, f\rangle \notin Z(\mathbf{B})$.

Now let α be a maximal member of the set of congruences χ of $\mathbf{B}$ such that $\langle e/\chi, f/\chi\rangle \notin Z(\mathbf{B}/\chi)$. Thus we can choose an $m+1$ -ary term s and

$\bar{z}, \bar{w} \in B^m$ such that, say

$$\langle s^{\mathbf{B}}(e, \bar{z}), s^{\mathbf{B}}(e, \bar{w}) \rangle \in \alpha \text{ and}$$

$$\langle s^{\mathbf{B}}(f, \bar{z}), s^{\mathbf{B}}(f, \bar{w}) \rangle \notin \alpha$$

It follows that the pair $\langle s^{\mathbf{B}}(f, \bar{z}), s^{\mathbf{B}}(f, \bar{w}) \rangle$ belongs to every congruence β such that $\alpha < \beta$ (by the maximality of α). Thus the algebra $\mathbf{F} = \mathbf{B}/\alpha$ is subdirectly irreducible. Letting $\hat{e} = e/\alpha$ and $\hat{f} = f/\alpha$, we have that $\langle \hat{e}, \hat{f} \rangle \notin Z(\mathbf{F})$ but $\langle \hat{e}/\beta, \hat{f}/\beta \rangle \in Z(\mathbf{F}/\beta)$ whenever β is a congruence of $\mathbf{F}$ such that $0_F < \beta$. Thus if $Z(\mathbf{F}) \neq 0_F$ then $\mathbf{F}/Z(\mathbf{F})$ is not centerless. By Theorem 2.1 we conclude that $\mathbf{F}$ is centerless. By Corollary 3.2, $\mathbf{F}$ has no no-zero Abelian congruence. Thus the type of the monolith of $\mathbf{F}$ cannot be **1** or **2**, implying that $\mathbf{F} \in \mathcal{V}_3$ by Definition 1.1. It is clear that $\mathbf{F} \not\models p(\bar{x}) \approx q(\bar{x})$ (since $e/\alpha \neq f/\alpha$). This contradicts our assumption that $\mathcal{V}_3 \models p(\bar{x}) \approx q(\bar{x})$. □

In proving Theorem 4.1, we shall require the two results below. Recall from §0.3 that for a class $\mathcal{K}$ of algebras, $\boldsymbol{P}_{\mathsf{fin}}(\mathcal{K})$ denotes the class of algebras isomorphic to a product $\mathbf{A}_1 \times \cdots \times \mathbf{A}_n$ for some non-negative integer n and some $\mathbf{A}_1, \ldots, \mathbf{A}_n \in \mathcal{K}$. Recall that Maltsev varieties were defined in the second paragraph of §0.5.

THEOREM 4.3 *Let $\mathcal{K}$ be a finite class of similar finite algebras. If $\boldsymbol{SP}_{\mathsf{fin}}(\mathcal{K}) \subseteq \boldsymbol{P}(\mathcal{K})$, then the variety generated by $\mathcal{K}$ is Maltsev.*

PROOF. This theorem is an immediate corollary of Theorem 3.4 in R. McKenzie [1982a]. □

THEOREM 4.4 *Let $\mathcal{W}$ be a locally finite Maltsev variety. If $\mathcal{W}$ is structured, then the class of finite, simple, non-Abelian algebras in $\mathcal{W}$ generates a discriminator variety.*

PROOF. If the locally finite variety $\mathcal{W}$ is the product of an affine variety $\mathcal{A}$ and a discriminator variety $\mathcal{D}$, then $\mathcal{D}$ is the variety generated by the class of finite, simple, non-Abelian algebras in $\mathcal{W}$. If $\mathcal{W}$ does not decompose in this way, then since $\mathcal{W}$ is Maltsev and hence congruence-modular, the arguments in Burris, McKenzie [1981] prove that either $\mathcal{W}$ interprets the class $\mathcal{BP}_1$ or $\mathcal{W}$ has a finite, centerless, subdirectly irreducible algebra with a nontrivial Abelian subalgebra. Then by Lemma 3.5 and the fact that $\mathcal{BP}_1$ is unstructured (see Theorem 0.47) it follows that $\mathcal{W}$ is unstructured. □

We shall also require two lemmas and a definition.

LEMMA 4.5 *$\mathcal{S}_3$ is identical with the class of finite, simple, non-Abelian algebras in $\mathcal{V}$. Every non-trivial subalgebra of an algebra in $\mathcal{S}_3$ belongs to $\mathcal{S}_3$.*

PROOF. This is an immediate consequence of Theorems 1.2 and 3.1. □

Definition 4.6 (1) A **subdirect product** of $\mathbf{B}_1, \ldots, \mathbf{B}_n$ is an algebra $\mathbf{A} \leq \mathbf{B}_1 \times \cdots \times \mathbf{B}_n$ such that $\mathbf{A}$ maps onto each of the algebras $\mathbf{B}_i$ via the coordinate projection.

(2) An **irredundant subdirect product** of $\mathbf{B}_1, \ldots \mathbf{B}_n$ is a subdirect product $\mathbf{A}$ of $\mathbf{B}_1, \ldots \mathbf{B}_n$ such that for each i, $1 \leq i \leq n$, $\mathbf{A}$ fails to be embedded in $\prod_{j \neq i} \mathbf{B}_j$ via the natural projection.

(3) A subdirect product $\mathbf{A}$ of $\mathbf{B}_1, \ldots, \mathbf{B}_n$ is called **direct** iff

$$\mathbf{A} = \mathbf{B}_1 \times \cdots \times \mathbf{B}_n.$$

The next lemma is the last result needed for proving Theorem 4.1.

LEMMA 4.7 *Every irredundant subdirect product of finitely many algebras in $\mathcal{S}_3$ is direct.*

PROOF. In order to derive a contradiction, we suppose that there exists an irredundant subdirect product of finitely many algebras in $\mathcal{S}_3$ which is not direct. We choose an algebra $\mathbf{A}$ which has this property and whose cardinality is least among all algebras having the property. Thus $\mathbf{A}$ is an irredundant subdirect product of $\mathbf{B}_1, \ldots, \mathbf{B}_n \in \mathcal{S}_3$, different from the direct product $\mathbf{B} = \mathbf{B}_1 \times \cdots \times \mathbf{B}_n$. Clearly $n > 1$. We are going to interpret in $\boldsymbol{P}_s(\mathbf{B})$ the class $\mathcal{BP}_1$ introduced in Definition 0.46. The proof runs more or less parallel to our proof of Theorem 0.48, but requires a considerably more complicated argument.

For each i, $1 \leq i \leq n$, denote by p_i the projection homomorphism of $\mathbf{B}$ onto $\mathbf{B}_i$, and by p_i' the projection homomorphism of $\mathbf{B}$ onto $\prod_{j \neq i} \mathbf{B}_j$. Observe that since $\mathbf{A}$ is an irredundant subdirect product, it follows for each i that $p_i'(\mathbf{A})$ is an irredundant subdirect product of the $\mathbf{B}_j$, $j \neq i$, and that $|p_i'(\mathbf{A})| < |\mathbf{A}|$. Thus since $\mathbf{A}$ was chosen to be a counterexample of minimum cardinality, we have

$$p_i'(A) = \prod_{j \neq i} B_j \text{ for each } i,\ 1 \leq i \leq n. \tag{4.1}$$

Denote by η_i the kernel of the homomorphism p_i, and by η_i' the kernel of p_i'. Then

η_i and η_i', $1 \le i \le n$, are congruences of $\mathbf{B}$ satisfying

$$\begin{array}{rl} \text{(i)} & \eta_i \text{ is a maximal congruence and } \mathbf{B}/\eta_i \cong \mathbf{B}_i \cong \mathbf{A}/(\eta_i|_A). \\ \text{(ii)} & \bigcap_i \eta_i = 0_B \text{ and } \bigcap_{j \ne i} \eta_j = \eta_i'. \\ \text{(iii)} & \eta_i'|_A \ne 0_A. \end{array} \tag{4.2}$$

Statement (i) follows from Lemma 4.5. (Each algebra $\mathbf{B}_i$ is a simple algebra.) Statement (ii) is obvious, and (iii) follows from the irredundance of the subdirect representation of $\mathbf{A}$.

Let $\mathbf{C} = \langle C_1, C_0, \le \rangle$ be a member of $\mathcal{BP}_1$ with $C_1 \subseteq \mathcal{P}(X)$. The algebra $\mathbf{D} \in \boldsymbol{P}_s(\mathbf{B})$ in which $\mathbf{C}$ will be interpreted can be defined immediately.

$\mathbf{D}' = \mathbf{A}[C_0, C_1, \eta_n'|_A]$ is the subalgebra of $\mathbf{A}^X$ consisting of all the functions f such that $f^{-1}(a) \in C_1$ and $f^{-1}(A \cap a/\eta_n') \in C_0$ for all $a \in A$.

$$\mathbf{D} \text{ is the ae-closure of } \mathbf{D}' \text{ in } \mathbf{B}^X. \tag{4.3}$$

It is easy to see that the mapping φ defined by $\varphi(f,g) = [\![f\eta_n g]\!]$ (where $[\![f\eta_n g]\!] = \{x : \langle f(x), g(x)\rangle \in \eta_n\}$) maps D^2 onto C_1: Using (4.2), let c_1 and c_2 be two elements of A such that $\langle c_1, c_2 \rangle \in \eta_n' - \eta_n$. For any $Y \in C_1$ the function f defined by $f(x) = c_1$ for $x \in Y$ and $f(x) = c_2$ for $x \notin Y$ belongs to $\mathbf{D}$; the constant function $g(x) = c_1$ belongs to $\mathbf{D}$; and $Y = \varphi(f,g)$.

Thus we hope to interpret elements of C_1 by equivalence classes of pairs of elements of D. To make this work, we need to find formulas defining in $\mathbf{D}$ the 4-ary relation

$$R_n(f,g,h,k) \ : \ [\![f\eta_n g]\!] \subseteq [\![h\eta_n k]\!], \tag{4.4}$$

and the binary relation

$$S_n(f,g) \ : \ [\![f\eta_n g]\!] \in C_0. \tag{4.5}$$

Note that the kernel of φ is definable if the relation R_n is definable. Our formulas can incorporate finitely many parameters that can be interpreted by any elements of $\mathbf{D}$, but the formulas must be independent of any special properties of the particular Boolean pair $\mathbf{C}$.

To find a first order definition of the relation R_n, we apply Lemma 3.3. Let $\mathrm{P}_{\subseteq}(x,y,u,v)$ and $\min(x,y)$ be the formulas supplied by that lemma for the collection $\{\mathbf{B}_1, \ldots, \mathbf{B}_n\}$. Define $Y = \{1, \ldots, n\} \times X$ and $\mathbf{B}_{\langle i,x\rangle} = \mathbf{B}_i$ for $\langle i,x\rangle \in Y$; and let $\mathbf{E}$ denote the subalgebra of $\prod_{y \in Y} \mathbf{B}_y$ corresponding to $\mathbf{D}$ under the natural isomorphism of this product with $\mathbf{B}^X$. Thus $\mathbf{E}$ is an ae-closed subdirect product of the $\mathbf{B}_y$. Let e_i be the element of $\mathbf{E}$ corresponding to $\hat{c}_i$ ($i = 1, 2$) under the isomorphism, where c_1 and c_2 are the two elements chosen previously.

Now let $x, y \in D$ and u, v be the corresponding elements of E. Since $\mathbf{E} \models \mathrm{P}_{\subseteq}(e_1, e_2, u, v)$ iff $[\![e_1 = e_2]\!] \subseteq [\![u = v]\!]$ (by Lemma 3.3), we see (via the isomorphism) that $\mathbf{D} \models \mathrm{P}_{\subseteq}(\hat{c}_1, \hat{c}_2, x, y)$ iff $[\![x \eta_n' y]\!] = X$. Further reasoning along these lines will establish that the relation $R_n(f, g, h, k)$ of (4.4) is defined in $\mathbf{D}$ by the formula

$$\mathrm{P}_n(x, y, u, v) \overset{\text{def}}{\longleftrightarrow}$$

$$(\forall z, w)\{(\mathrm{P}_{\subseteq}(\hat{c}_1, \hat{c}_2, z, w) \wedge \mathrm{P}_{\subseteq}(u, v, z, w)) \longrightarrow \mathrm{P}_{\subseteq}(x, y, z, w)\}. \tag{4.6}$$

Since R_n is defined by the formula P_n, the formula $\Delta_n(x, y) \equiv \mathrm{P}_n(x, x, x, y)$ has the property

$$\mathbf{D} \models \Delta_n(f, g) \longleftrightarrow (\forall x)(f(x)\, \eta_n\, g(x)). \tag{4.7}$$

Moreover, we already observed that the formula $\Delta_n'(x, y) \equiv \mathrm{P}_{\subseteq}(\hat{c}_1, \hat{c}_2, x, y)$ has the property

$$\mathbf{D} \models \Delta_n'(f, g) \longleftrightarrow (\forall x)(f(x) \eta_n' g(x)). \tag{4.8}$$

To construct a first order definition of the relation S_n (of (4.5)), we need to choose three special elements of $\mathbf{A}$ satisfying

$$\begin{gathered} a_0, a_1, a_2 \in A, \langle a_0, a_1 \rangle \in \eta_n' - \eta_n, \\ \text{and the element } u \in B \text{ such that } a_1\ \eta_n\ u\ \eta_n'\ a_2 \\ \text{does not belong to } A. \end{gathered} \tag{4.9}$$

To find such elements, we first choose some element $u \in B - A$, and let $c_1, c_2 \in A$ be the elements previously chosen. Then we choose $d \in A$ with $d \eta_n u$. Since $\langle c_1, c_2 \rangle \in \eta_n' - \eta_n$, one of c_1 and c_2 fails to be η_n-equivalent to d; we assume that $\langle c_1, d \rangle \notin \eta_n$. Let $v \in B$ be the element such that $c_1 \eta_n v \eta_n' d$. If $v \notin A$ then we take $a_0 = c_2$, $a_1 = c_1$, $a_2 = d$. If $v \in A$ then we take $a_0 = v$, $a_1 = d$, and let a_2 be some element of A such that $a_2 \eta_n' u$. (The existence of a_2 follows from (4.1)).

Continuing to work toward a first order definition of S_n, we now show that the relation of almost nowhere η_n-equivalence in D is definable.

$$\begin{gathered} \text{For all } f, g \in D, [\![f \eta_n g]\!] \text{ is finite iff there exists} \\ h \in D \text{ such that } \mathbf{D} \models \Delta_n'(h, \hat{a}_2) \text{ and } [\![f \eta_n g]\!] = [\![h \eta_n \hat{a}_1]\!]. \end{gathered} \tag{4.10}$$

To prove (4.10), let u be the element such that $a_1 \eta_n u \eta_n' a_2$. By (4.9), $u \in B - A$. Suppose that $[\![f \eta_n g]\!]$ is a finite set. Define $h \in B^X$ to be the function such that $h(x) = u$ when $f(x)\eta_n g(x)$ and $h(x) = a_2$ otherwise. Since $h \overset{\text{ae}}{=} \hat{a}_2$, $h \in D$. Clearly, $\mathbf{D} \models \Delta_n'(h, \hat{a}_2)$ and $[\![h\eta_n \hat{a}_1]\!] = [\![f\eta_n g]\!]$. (Observe that $a_2\eta_n a_1$ would imply $a_2 = u$.) Conversely, suppose that $f, g, h \in D$, that $\mathbf{D} \models \Delta_n'(h, \hat{a}_2)$, and that $[\![f\eta_n g]\!] = [\![h\eta_n \hat{a}_1]\!]$. Now if $x \in [\![h\eta_n \hat{a}_1]\!]$ then $a_2\eta_n' h(x)\eta_n a_1$, implying that $h(x) = u$. But since $h \overset{\text{ae}}{=} h'$ for some $h' \in D' \subseteq A^X$, the set $h^{-1}(u)$ must be finite. Thus $[\![f\eta_n g]\!]$ is a finite set. This concludes our argument for (4.10).

Since $\{[\![f\eta_n g]\!] : f, g \in D\} = C_1$ is a Boolean algebra of subsets of X and the relation R_n (of (4.4)) is definable in $\mathbf{D}$, it follows easily from (4.10) that there are formulas $\mathrm{A}_n(x, y, u, v)$ and $\mathrm{M}_n(x, y, u, v)$ such that

$$\mathbf{D} \models \mathrm{A}_n(f, g, h, k) \text{ iff } [\![f\eta_n g]\!] \cap [\![h\eta_n k]\!] \text{ is finite;} \tag{4.11}$$

$$\mathbf{D} \models \mathrm{M}_n(f, g, h, k) \text{ iff } [\![f\eta_n g]\!] \cup [\![h\eta_n k]\!] = X. \tag{4.12}$$

We are now in a position to conclude the proof of this Lemma by showing that the relation

$$S_n(f, g) : \ [\![f\eta_n g]\!] \in C_0$$

is definable in $\mathbf{D}$.

Claim: If $f, g \in D$ then $[\![f\eta_n g]\!] \in C_0$ iff there exist $h, k \in D$ such that $[\![f\eta_n g]\!] = [\![h\eta_n k]\!]$ and

(1) $\mathbf{D} \models \Delta_n'(h, k)$;

(2) $\mathrm{M}_n(h, k, h, \hat{a}_0) \wedge \mathrm{M}_n(h, k, k, \hat{a}_1)$;

(3) $(\forall \alpha \in D)(\mathbf{D} \models \Delta_n'(\alpha, h) \to \mathrm{A}_n(\alpha, \hat{a}_1, h, k))$.

To prove this assertion, suppose first that $[\![f\eta_n g]\!] = Y \in C_0$. Define h and k so that $h(x) = k(x) = a_2$ for $x \in Y$, while $h(x) = a_0$ and $k(x) = a_1$ for $x \in X - Y$. Clearly the functions h and k belong to D'. Since a_0 and a_1 are incongruent modulo η_n, we have $[\![f\eta_n g]\!] = [\![h\eta_n k]\!]$. The condition (1) holds because $a_0\eta_n' a_1$. Then (2) is obvious from the definition of M_n and of the functions h and k. To prove (3), suppose that $\alpha \in D$ and $\alpha(x)\eta_n' h(x)$ for all $x \in X$. If x is such that $h(x)\eta_n k(x)$ and $\alpha(x)\eta_n a_1$ then $h(x) = a_2$ and $a_2\eta_n'\alpha(x)\eta_n a_1$, implying that $\alpha(x) \notin A$ by (4.9). Since $\alpha \overset{\text{ae}}{=} \alpha'$ for some $\alpha' \in A^X$, there can be only a finite number of $x \in X$ for which these conditions are satisfied. Thus (3) is established.

Conversely, suppose that $f, g, h, k \in D$ satisfy (1)–(3) and that $[\![f\eta_n g]\!] = [\![h\eta_n k]\!] = Y$. We need to show that $Y \in C_0$. Condition (1) implies that $h(x) = k(x)$ iff $h(x)\eta_n k(x)$; thus $Y = [\![h = k]\!]$. Condition (2) implies that for $x \notin Y$, $h(x)\eta_n a_0$ and $k(x)\eta_n a_1$. We claim that (3) implies that for every

$a \in A$, either $[\![h\eta_n{}'\hat{a}]\!] \cap Y$ is finite or $[\![h\eta_n{}'\hat{a}]\!] - Y$ is finite. Suppose that the claim is false, for a certain $a \in A$. For $x \in [\![h\eta_n{}'\hat{a}]\!] - Y$ we have that $k(x)$ is the unique element $b \in B$ satisfying $a_1 \eta_n b \eta_n{}' a$. Since there are infinitely many such x, it follows that this element b belongs to A. Let $\alpha \in B^X$ be the function such that $\alpha(x) = b$ when $h(x)\eta_n{}'a$, and $\alpha(x) = k(x)$ otherwise. It is easy to check that $\alpha \in D$ and $\mathbf{D} \models \Delta_n{}'(\alpha, h)$ (using that $h, k \in D$ and $b \in A$ and $\mathbf{D} \models \Delta_n{}'(h, k)$). Moreover, the infinite set $[\![h\eta_n{}'\hat{a}]\!] \cap Y$ is included in $[\![h\eta_n k]\!] \cap [\![\alpha\eta_n \hat{a}_1]\!]$. This contradicts (3).

Now the sets $Z_a = [\![h\eta_n{}'\hat{a}]\!]$, where a ranges over a set of representatives of the $\eta_n{}'|_A$-equivalence classes, partition X. Each of the sets Z_a belongs to C_0. We have

$$Y = \bigcup_a (Y \cap Z_a);$$

and by the claim, each of the sets $Y \cap Z_a$ is finite or almost equal to Z_a. Since the Boolean algebra C_0 includes all of the finite subsets of X, it now follows that $Y \in C_0$. This finishes our proof of the claim, and with it the proof of Lemma 4.7. □

Proof of Theorem 4.1. By Lemma 4.5, $\mathcal{V}_3$ is generated by the class of finite, simple, non-Abelian algebras in $\mathcal{V}_3$. Thus if $\mathcal{V}_3$ can be shown to be Maltsev, then it will follow by Theorem 4.4 that $\mathcal{V}_3$ is a discriminator variety. We will apply Theorem 4.3 and Lemma 4.7 to prove that $\mathcal{V}_3$ is Maltsev.

Let $\mathbf{F}$ be the free algebra on three generators in $\mathcal{V}_3$. Thus $\mathcal{V}_3$ is Maltsev iff $\mathbf{F}$ has permuting congruences. Since $\mathbf{F}$ is finite, and $\mathcal{V}_3 = \boldsymbol{HSP}(\mathcal{S}_3)$ is locally finite, there exists a finite set $\mathcal{K} \subseteq \mathcal{S}_3$ such that $\mathbf{F} \in \boldsymbol{HSP}(\mathcal{K})$. Thus our task is reduced to proving that $\mathcal{V}' = \boldsymbol{HSP}(\mathcal{K})$ is Maltsev. By Lemma 4.5 (second statement), we can assume that every nontrivial subalgebra of an algebra in $\mathcal{K}$ belongs to $\mathcal{K}$. Then $\boldsymbol{SP}_{\text{fin}}(\mathcal{K})$ consists of the trivial one-element algebras and the algebras isomorphic to irredundant subdirect products of finitely many members of $\mathcal{K}$. By Lemma 4.7, we have $\boldsymbol{SP}_{\text{fin}}(\mathcal{K}) \subseteq \boldsymbol{P}(\mathcal{K})$. Thus by Theorem 4.3, $\mathcal{V}'$ is Maltsev. □

Chapter 5

The Abelian subvariety

We continue to work with a fixed locally finite variety that is structured. In this chapter we introduce certain properties of finite algebras that we call transfer principles. Using a theorem concerning these principles which will be proved in the next chapter, we give a short proof that $\mathcal{V}_1 \vee \mathcal{V}_2$ is an Abelian variety. The concept which we now introduce, and our reasoning about it, depend heavily on tame congruence theory. (See §0.6.)

Definition 5.1 Let $\mathbf{A}$ be a finite algebra. If $\mathbf{i},\mathbf{j} \in \{\mathbf{1},\ldots,\mathbf{5}\}$ are distinct types, we say that $\mathbf{A}$ possesses the $(\mathbf{i},\mathbf{j})$ **transfer principle** iff for all $\chi_0,\chi_1,\chi_2 \in \mathrm{Con}\,\mathbf{A}$, if $\chi_0 \prec \chi_1 \prec \chi_2$ and $\mathrm{typ}(\chi_0,\chi_1) = \mathbf{i}$ and $\mathrm{typ}(\chi_1,\chi_2) = \mathbf{j}$, then there exists $\beta \le \chi_2$ such that $\chi_0 \prec \beta$ and $\mathrm{typ}(\chi_0,\beta) = \mathbf{j}$.

Definition 5.2 Let $\mathbf{A}$ be a finite algebra. If $\chi_0 < \chi_1 < \chi_2$ are congruences of $\mathbf{A}$ such that χ_1 is the only congruence β satisfying $\chi_0 < \beta < \chi_2$, then the three-element interval $I[\chi_0,\chi_2]$ in $\mathrm{Con}\,\mathbf{A}$ will be said to have type $(\mathbf{i},\mathbf{j})$ where $\mathbf{i} = \mathrm{typ}(\chi_0,\chi_1)$ and $\mathbf{j} = \mathrm{typ}(\chi_1,\chi_2)$.

LEMMA 5.3 *A finite algebra* $\mathbf{A}$ *possesses the* $(\mathbf{i},\mathbf{j})$ *transfer principle iff* $\mathbf{Con\,A}$ *has no three-element interval of type* $(\mathbf{i},\mathbf{j})$.

PROOF. Suppose that $\chi_0 \prec \chi_1 \prec \chi_2$ in $\mathbf{Con\,A}$ and that $I[\chi_0,\chi_2]$ contains a congruence α different from χ_0,χ_1 and χ_2. Let β be a minimal member of the interval $I[\chi_0,\alpha]$. Now every member of $I[\chi_0,\chi_2]$ different from χ_0,χ_1 and χ_2 is incomparable to χ_1, since $\chi_0 \prec \chi_1 \prec \chi_2$. Thus it follows easily that β is incomparable to χ_1. Then $\beta \vee \chi_1 = \chi_2$ and $\beta \wedge \chi_1 = \chi_0$ as there are no other possibilities for this join and meet. By Lemma 0.37, we have $\mathrm{typ}(\chi_0,\beta) = \mathrm{typ}(\chi_1,\chi_2)$. The lemma follows immediately from these observations. □

Chapter 6 contains a proof that every finite algebra in $\mathcal{V}$ possesses the $(\mathbf{1},\mathbf{3})$ and the $(\mathbf{2},\mathbf{3})$ transfer principles (Theorem 6.1). Using this result, we can prove the second principal theorem of Part I, which follows.

THEOREM 5.4 *$\mathcal{V}_1 \vee \mathcal{V}_2$ is the class of all Abelian algebras in $\mathcal{V}$.*

PROOF. By Theorem 1.3, every Abelian algebra in $\mathcal{V}$ belongs to $\mathcal{V}_1 \vee \mathcal{V}_2$. To prove the converse, it will suffice, by Corollary 3.2, to prove that every algebra in $\mathcal{V}_1 \vee \mathcal{V}_2$ is locally solvable. Since $\mathcal{V}_1 \vee \mathcal{V}_2$ is generated by the class $\mathcal{S}_1 \cup \mathcal{S}_2$ (Definition 1.1), it will suffice, by Theorem 0.15, to prove that every algebra in $\mathcal{S}_1 \cup \mathcal{S}_2$ is solvable. Now $\mathcal{S}_1 \cup \mathcal{S}_2$ is just the class of finite subdirectly irreducible algebras in $\mathcal{V}$ with Abelian monolith. So let $\mathbf{A}$ be a finite subdirectly irreducible algebra belonging to $\mathcal{V}$ whose monolith β is Abelian. We shall show that $\mathbf{A}$ is solvable.

By Theorem 0.39, we have $\text{typ}\{\mathbf{A}\} \subseteq \{\mathbf{1},\mathbf{2},\mathbf{3}\}$. By Theorem 0.36(iii), $\mathbf{A}$ is solvable iff $\text{typ}\{\mathbf{A}\} \subseteq \{\mathbf{1},\mathbf{2}\}$. Now in order to reach a contradiction, assume that $\mathbf{3} \in \text{typ}\{\mathbf{A}\}$. Among all the prime quotients $\langle \chi_0, \chi_1 \rangle$ in $\mathbf{Con}\,\mathbf{A}$ with $\text{typ}(\chi_0, \chi_1) = \mathbf{3}$, choose one $\langle \delta, \gamma \rangle$ such that the cardinality of the interval $I[0_A, \gamma]$ is as small as possible. Since $\text{typ}(0_A, \beta) \in \{\mathbf{1},\mathbf{2}\}$ where β is the monolith, we have $\delta \neq 0_A$. Now choose any $\xi \in \text{Con}\,\mathbf{A}$ such that $\xi \prec \delta$. By our choice of $\langle \delta, \gamma \rangle$, the type of $\langle \xi, \delta \rangle$ is not $\mathbf{3}$—so it must be $\mathbf{1}$ or $\mathbf{2}$. Now by the $(\mathbf{1},\mathbf{3})$ or the $(\mathbf{2},\mathbf{3})$ transfer principle (see Theorem 6.1), there exists a congruence λ such that $\xi \prec \lambda < \gamma$ and $\text{typ}(\xi, \lambda) = \mathbf{3}$. This contradicts our choice of $\langle \delta, \gamma \rangle$ and ends the proof. □

Chapter 6

Transfer principles

We continue to assume that $\mathcal{V}$ is a structured locally finite variety. This chapter is devoted to a proof of the theorem below, on which our proof of Theorem 5.4 was based.

THEOREM 6.1 *Every finite algebra in $\mathcal{V}$ possesses the* $(\mathbf{1},\mathbf{3})$ *and* $(\mathbf{2},\mathbf{3})$ *transfer principles.*

The transfer principles were defined at the beginning of Chapter 5. If the theorem were not true then, according to Definition 5.2 and Lemma 5.3, $\mathcal{V}$ would have a finite algebra $\mathbf{A}$ with a three-element interval $I[0_A,\beta] = \{0_A,\alpha,\beta\}$ in $\mathbf{Con\ A}$ of type $(\mathbf{1},\mathbf{3})$ or $(\mathbf{2},\mathbf{3})$. Then, choosing any N in $M_{\mathbf{A}}(\alpha,\beta)$, we would have by Theorem 0.39 that N is a two-element set, say $N = \{0,1\}$, and that $\mathbf{A}|_N$ is polynomially equivalent with the two-element Boolean algebra. The proof of the theorem then divides into cases, depending on whether or not there exists a polynomial function f of $\mathbf{A}$ such that $\langle f(0), f(1)\rangle \in \alpha - 0_A$; and if not, whether $\mathrm{typ}(0_A,\alpha)$ is $\mathbf{1}$ or $\mathbf{2}$. We can easily dispose of the last of these three cases.

LEMMA 6.2 *Let $\mathbf{A}$ be a finite algebra in $\mathcal{V}$. Suppose that $I[0_A,\beta]$ is a three-element interval in* $\mathbf{Con\ A}$ *of type* $(\mathbf{2},\mathbf{3})$*, where $0_A \prec \alpha \prec \beta$. If $N = \{0,1\}$ belongs to $M_{\mathbf{A}}(\alpha,\beta)$ then there exists $f \in \mathrm{Pol}_1\mathbf{A}$ such that $\langle f(0), f(1)\rangle \in \alpha - 0_A$.*

PROOF. Let $N = \{0,1\} \in M_{\mathbf{A}}(\alpha,\beta)$ and $M \in M_{\mathbf{A}}(0_A,\alpha)$. According to Theorem 0.39, $\mathbf{A}|_M$ is Maltsev. Let $e \in \mathrm{E}(\mathbf{A})$ satisfy $e(A) = M$. The congruence $\theta = \mathrm{Cg}^{\mathbf{A}}(0,1)$ satisfies $\alpha \leq \theta$ since $I[0_A,\beta]$ is a three-element interval. Thus letting $\langle a,b\rangle \in \alpha|_M$, $a \neq b$, there must exist polynomial functions $f_0, \ldots, f_m$ such that

$$a = f_0(0), b = f_m(1) \text{ and for all } i < m$$
$$f_i(1) = f_{i+1}(0).$$

(Notice that since $\mathbf{A}|_N$ is a two-element algebra polynomially equivalent to a Boolean algebra, there exists a polynomial function $c(x)$ such that $c(0) = 1$ and $c(1) = 0$; whence the pairs of elements in the congruence θ can be characterized in this relatively simple way.) Applying the operation e to all of the elements in the displayed formula, and noting that $a = e(a)$ and $b = e(b)$, we find that $\langle a, b\rangle$ belongs to the congruence on $\mathbf{A}|_M$ generated by the set $T = \{\langle ef_i(0), ef_i(1)\rangle : i \leq m\}$. Since $\mathbf{A}|_M$ is Maltsev, there exists a polynomial operation $p(x_0, \ldots, x_m)$ of $\mathbf{A}$ such that

$$a = p(ef_0(0), \ldots, ef_m(0)) \text{ and}$$
$$b = p(ef_0(1), \ldots, ef_m(1)).$$

(See Theorem 0.20.) Then where $f(x) = p(ef_0(x), \ldots, ef_m(x))$ we have that $\langle f(0), f(1)\rangle \in \alpha - 0_A$ as desired. □

The next two lemmas dispose of the analogous case when $\operatorname{typ}(0_A, \alpha)$ is **1**. Recall that for congruences δ and γ of an algebra, $(\delta : \gamma)$ denotes the largest congruence μ such that μ centralizes γ modulo δ. (See §0.4.)

LEMMA 6.3 *Let* $\mathbf{A}$ *be a finite algebra in* $\mathcal{V}$. *Suppose that* α *and* β *are congruences of* $\mathbf{A}$ *such that* α *is Abelian,* $\alpha \prec \beta$, $\operatorname{typ}(\alpha, \beta) = \mathbf{3}$, *and* $\beta \not\leq (0_A : Z(\mathbf{A}))$. *If* $N = \{0, 1\}$ *belongs to* $M_{\mathbf{A}}(\alpha, \beta)$, *then there exists* $f \in \operatorname{Pol}_1 \mathbf{A}$ *such that* $\langle f(0), f(1)\rangle \in \alpha - 0_A$.

PROOF. Let $N = \{0, 1\} \in M_{\mathbf{A}}(\alpha, \beta)$. By Corollary 3.2, $\alpha \leq Z(\mathbf{A}) \leq (0_A : Z(\mathbf{A}))$. Since $\alpha \prec \beta$ we have that $\beta = \alpha \vee \mathrm{Cg}^{\mathbf{A}}(0, 1)$; and thus since $\beta \not\leq (0_A : Z(\mathbf{A}))$, we have that $\langle 0, 1\rangle \notin (0_A : Z(\mathbf{A}))$. Notice that $\beta \wedge Z(\mathbf{A}) = \alpha$. Therefore, if this lemma is false, then we have

$$\neg(\exists f \in \operatorname{Pol}_1 \mathbf{A})(\langle f(0), f(1)\rangle \in Z(\mathbf{A}) - 0_A). \tag{6.1}$$

Let us assume that (6.1) holds. We shall interpret the class $\mathcal{BP}_1$ of Definition 0.46 into $\boldsymbol{P}_s(\mathbf{A})$, contradicting our assumption that $\mathcal{V}$ is structured.

Let $\mathbf{B} = \langle B_1, B_0, \leq\rangle \in \mathcal{BP}_1$ with $B_1 \subseteq \mathcal{P}(X)$. We define an algebra $\mathbf{D} \subseteq \mathbf{A}^X$. Recall that $U[B_i]$ denotes the set of all functions $f \in U^X$ such that $f^{-1}(u) \in B_i$ for all $u \in U$. Let Δ be the set of functions $f \in A[B_0]$ such that $\langle f(x), f(y)\rangle \in Z(\mathbf{A})$ for all $x, y \in X$.

$$\mathbf{D}' \text{ is the subalgebra of } \mathbf{A}^X \text{ generated by } N[B_1] \cup \Delta. \tag{6.2}$$

$$\mathbf{D} \text{ is the ae-closure of } \mathbf{D}' \text{ in } \mathbf{A}^X.$$

Observe that $\Delta \subseteq \mathbf{A}[B_1]$ and so $\mathbf{D} \subseteq \mathbf{A}[B_1]$. Also, Δ contains all the constant functions, so $\mathbf{D}$ is an ae-closed diagonal subdirect power of $\mathbf{A}$. We shall build an interpretation of $\mathbf{B}$ into $\mathbf{D}$.

By Theorem 0.25 (ii), there exists $e \in \mathrm{E}(\mathbf{A})$ with $e(A) = N$. We let $\bar{a} = \langle a_0, \ldots, a_{k-1}\rangle$ where $A = \{a_0, \ldots, a_{k-1}\}$. Then we choose a term $e(x, \bar{z})$ such that $e^{\mathbf{A}}(x, \bar{a}) = e(x)$. Since $\mathbf{A}|_N$ is a minimal algebra of type **3**, we can choose terms $s(x, y, \bar{z})$, $m(x, y, \bar{z})$ and $c(x, \bar{z})$ such that $s^{\mathbf{A}}(x, y, \bar{a})$ and $m^{\mathbf{A}}(x, y, \bar{a})$, restricted to N, are the Boolean join and meet, and $c^{\mathbf{A}}(x, \bar{a})$ restricted to N is the Boolean complementation. Taking $D(N) = D \cap N^X$ as usual, it is clear that $D(N) = N[B_1]$ and that $D(N)$ is definable in $\mathbf{D}$ as the range of the polynomial function $e^{\mathbf{D}}(x, \hat{a}_0, \ldots, \hat{a}_{k-1})$, which is just $e^{(X)}|_D$ (i.e., e acting coordinatewise). The natural order on $D(N)$ is definable in $\mathbf{D}$, in fact we have

$$f \leq g \longleftrightarrow (\{f, g\} \subseteq D(N) \wedge f = m^{\mathbf{D}}(f, g, \hat{a}_0, \ldots, \hat{a}_{k-1})).$$

Now the definable structure $\langle D(N), \leq\rangle$ is isomorphic to $\langle B_1, \leq\rangle$ by the mapping $f \mapsto f^{-1}(1)$. To complete our interpretation of $\mathbf{B}$ into $\mathbf{D}$, we only require a first order formula to define the set $N[B_0]$ $(= \{f \in D(N) : f^{-1}(1) \in B_0\})$.

Recall that $\langle 0, 1\rangle \notin (0_A : Z(\mathbf{A}))$. This means that there exists an $n+1$-ary term $p(x, \bar{z})$ (for some n) and $\bar{c}, \bar{d} \in A^n$ with $\bar{c} \equiv \bar{d} \pmod{Z(\mathbf{A})}$ (i.e., $\langle c_i, d_i\rangle \in Z(\mathbf{A})$ for each $i < n$), such that

$$p^{\mathbf{A}}(0, \bar{c}) = p^{\mathbf{A}}(0, \bar{d}) \text{ and } p^{\mathbf{A}}(1, \bar{c}) \neq p^{\mathbf{A}}(1, \bar{d}). \tag{6.3}$$

(Or the same condition holds with 0 and 1 interchanged. But since there exists a polynomial function which exchanges 0 and 1, we get (6.3) in any case.) Putting $E(\bar{u}, \bar{v}) = \{x : p(x, \bar{u}) = p(x, \bar{v})\}$, we have that $0 \in E^{\mathbf{A}}(\bar{c}, \bar{d})$ and $1 \notin E^{\mathbf{A}}(\bar{c}, \bar{d})$.

Let $\mathrm{CEN}_1(x, y)$ be the formula from the proof of Lemma 2.3; so that

$$\mathbf{D} \models \mathrm{CEN}_1(f, g) \text{ iff } (\forall x \in X)(\langle f(x), g(x)\rangle \in Z(\mathbf{A})). \tag{6.4}$$

Let $\bar{c}' = \langle \hat{c}_0, \ldots, \hat{c}_{n-1}\rangle$ and $\bar{d}' = \langle \hat{d}_0, \ldots, \hat{d}_{n-1}\rangle$. Then we claim that for all $f \in D(N)$

$$\begin{array}{l} f \in N[B_0] \text{ iff there exist } g_0, \ldots, g_{n-1} \in D \text{ such that} \\ \mathbf{D} \models \mathrm{CEN}_1(g_i, \hat{c}_i) \text{ for each } i < n,\ \hat{0} \in E^{\mathbf{D}}(\bar{c}', \bar{g}), \\ \text{and } f \text{ is the largest member of } D(N) \cap E^{\mathbf{D}}(\bar{c}', \bar{g}). \end{array} \tag{6.5}$$

The proof of this claim will finish the proof of this lemma.

To prove (6.5), suppose first that $f \in N[B_0]$. Define $g_0, \ldots, g_{n-1}$ so that $g_i(x) = d_i$ when $f(x) = 0$ and $g_i(x) = c_i$ otherwise. Clearly, these functions belong to Δ, and so belong to $\mathbf{D}$; moreover $\mathbf{D} \models \mathrm{CEN}_1(g_i, \hat{c}_i)$ since $\langle d_i, c_i \rangle \in Z(\mathbf{A})$. From our definition of the g_i, and from (6.3), a function $h \in D(N)$ belongs to $E^{\mathbf{D}}(\bar{c}', \bar{g})$ iff $[\![f = 0]\!] \subseteq [\![h = 0]\!]$, or what amounts to the same thing, iff $h \leq f$. Thus the condition in (6.5) is a necessary condition for $f \in N[B_0]$.

To prove that the condition is sufficient, suppose that $f \in D(N)$, and that $g_0, \ldots, g_{n-1} \in D$ are such that $\langle g_i(x), c_i \rangle \in Z(\mathbf{A})$ for all $x \in X$, $\hat{0} \in E^{\mathbf{D}}(\bar{c}', \bar{g})$, and f is the largest member of $D(N) \cap E^{\mathbf{D}}(\bar{c}', \bar{g})$. We first establish that $\{g_0, \ldots, g_{n-1}\} \subseteq \Delta$. Since $g_i(x) \equiv c_i \equiv g_i(y) \pmod{Z(\mathbf{A})}$ for all x, y, we only need to show that $g_i \in A[B_0]$. Here is the proof that $g_0 \in A[B_0]$. Of course, we have $g_0 \stackrel{\text{ae}}{=} g_0'$ for some $g_0' \in D'$. Then the set

$$Y = \{x \in X : \langle g_0'(x), c_0 \rangle \in Z(\mathbf{A})\}$$

contains all but finitely many elements of X. By (6.2) we can write

$$g_0' = H^{\mathbf{D}}(\bar{h}, \bar{k})$$

where H is a term, $\bar{h} \in (N[B_1])^m$, and $\bar{k} \in \Delta^l$ (for some m, l). For any $u, v \in Y$, since $g_0'(u) \equiv g_0'(v) \pmod{Z(\mathbf{A})}$ and $\bar{k}(u) \equiv \bar{k}(v) \pmod{Z(\mathbf{A})}$, it follows that

$$H^{\mathbf{A}}(\bar{h}(u), \bar{k}(u)) \equiv H^{\mathbf{A}}(\bar{h}(v), \bar{k}(u)) \pmod{Z(\mathbf{A})}.$$

Since $\bar{h}(u), \bar{h}(v) \in \{0,1\}^m$ and $\mathbf{A}|_N$ is a Boolean algebra, there are unary polynomial functions $r_0, \ldots, r_{m-1}$ of $\mathbf{A}$ such that

$$\bar{h}(u) = \bar{r}(0) \text{ and } \bar{h}(v) = \bar{r}(1).$$

Then the polynomial function

$$R(x) = H^{\mathbf{A}}(\bar{r}(x), \bar{k}(u))$$

satisfies

$$R(0) = H^{\mathbf{A}}(\bar{h}(u), \bar{k}(u)),\ R(1) = H^{\mathbf{A}}(\bar{h}(v), \bar{k}(u))$$
$$\langle R(0), R(1) \rangle \in Z(\mathbf{A}).$$

Now our assumption (6.1) implies that $R(0) = R(1)$, i.e., that $g_0'(u) = H^{\mathbf{A}}(\bar{h}(v), \bar{k}(u))$. From what we just proved, it follows that for all $u \in Y$, $g_0'(u) = k(u)$ where

$$k = H^{\mathbf{D}}(\epsilon_0, \ldots, \epsilon_{m-1}, \bar{k})$$

and the functions $\epsilon_0, \ldots, \epsilon_{m-1} \in D$ are constant. Clearly, $k \in A[B_0]$ because $\bar{k} \in \Delta^l$. We also have $g_0 \stackrel{\text{ae}}{=} g_0' \stackrel{\text{ae}}{=} k$. Thus $g_0 \in A[B_0]$ as we claimed. ($A[B_0]$ is ae-closed, because B_0 contains every finite subset of X.)

Now that we have shown that $g_0, \ldots, g_{n-1} \in \Delta$, we define

$$Z = \{x \in X : 1 \in E^{\mathbf{A}}(\bar{c}'(x), \bar{g}(x))\}.$$

Notice that $0 \in E^{\mathbf{A}}(\bar{c}'(x), \bar{g}(x))$ for all x. Then it follows that for any $h \in D(N)$, we have $h \in E^{\mathbf{D}}(\bar{c}', \bar{g})$ iff $[\![h = 1]\!] \subseteq Z$. Thus f is the largest element h of $D(N)$ with $[\![h = 1]\!] \subseteq Z$. It follows that $f^{-1}(1) = Z$ since $D(N)$ is ae-closed. All that remains is to observe that $Z \in B_0$. This follows from the fact that $\hat{c}_0, \ldots \hat{c}_{n-1}, g_0, \ldots, g_{n-1} \in A[B_0]$. □

The proof of the next lemma is quite a bit longer than the other proofs in this chapter.

LEMMA 6.4 *Let $\mathbf{A}$ be a finite algebra in $\mathcal{V}$. Suppose that $I[0_A, \beta]$ is a three-element interval in* **Con A** *of type* $(\mathbf{1}, \mathbf{3})$, *where $0_A \prec \alpha \prec \beta$. If $N = \{0,1\}$ belongs to $M_{\mathbf{A}}(\alpha, \beta)$ then there exists $f \in \mathrm{Pol}_1\mathbf{A}$ such that $\langle f(0), f(1)\rangle \in \alpha - 0_A$.*

PROOF. Let us assume that there does not exist $f \in \mathrm{Pol}_1\mathbf{A}$ such that $\langle f(0), f(1)\rangle \in \alpha - 0_A$. Then we shall interpret the class of of all graphs into $\boldsymbol{P}_s(\mathbf{A})$, contradicting our assumption that $\mathcal{V}$ is structured.

Let $M \in M_{\mathbf{A}}(0_A, \alpha)$ and say $M = e(A)$ with $e \in \mathrm{E}(\mathbf{A})$. Just as in the previous lemma, we have $\alpha \leq \mathrm{Cg}^{\mathbf{A}}(0,1)$; also, by that lemma, we have

$$\beta \leq (0_A : Z(\mathbf{A})). \tag{6.6}$$

By an argument used just above, we have that $\alpha|_M$ is contained in the transitive closure of the relation $\{\langle ef(0), ef(1)\rangle : f \in \mathrm{Pol}_1\mathbf{A}\}$. Thus there must exist a pair $\langle a, b\rangle \in \alpha|_M - 0_M$ and a polynomial function f $(= ef)$ such that $f(0) = a$ and $f(1) \in M - \{a\}$. From our initial assumption, we have that $\langle f(0), f(1)\rangle \in \beta - \alpha$. Then by Theorem 0.25 (iii), $\{f(0), f(1)\} \in M_{\mathbf{A}}(\alpha, \beta)$. Changing notation (replacing 0 by $f(0)$ and 1 by $f(1)$ and writing $0'$ for b) we have

$$\begin{gathered}\{0,1\} = N \subseteq M\,,\ M \in M_{\mathbf{A}}(0_A, \alpha)\,,\ N \in M_{\mathbf{A}}(\alpha, \beta), \\ 0' \in M \text{ and } \langle 0, 0'\rangle \in \alpha - 0_A.\end{gathered} \tag{6.7}$$

(This change obviously does not affect our assumption that there does not exist a polynomial function f such that $\langle f(0), f(1)\rangle \in \alpha - 0_A$.)

We let $A = \{a_0, \ldots, a_{k-1}\}$ and find some terms $t_0, \ldots, t_{r-1}$ so that the polynomial function $e_i(x) = t_i^{\mathbf{A}}(x, \bar{a})$ belongs to $\mathrm{E}(\mathbf{A})$ for each $i < r$, and so that

$$\begin{array}{c} M_{\mathbf{A}}(\alpha, \beta) = \{N_0, \ldots, N_{r-1}\} \text{ where} \\ N_i = e_i(A) \text{ and } N_0 = N. \end{array} \tag{6.8}$$

We choose also a term $t(x, \bar{z})$ such that $e(x) = t^{\mathbf{A}}(x, \bar{a})$ belongs to $\mathrm{E}(\mathbf{A})$ and $e(A) = M$.

Letting $\mathbf{W} = \langle W, E\rangle$ be any graph, we proceed to construct an algebra in which we can interpret $\mathbf{W}$. We can assume that W, E and $\{0,1\}^W$ are pairwise disjoint sets. We choose two different elements x_0 and x_1 outside the union of these sets, and put

$$X = \{x_0, x_1\} \cup W \cup E \cup \{0,1\}^W. \tag{6.9}$$

For each $w \in W$ we define a function $\mu_w \in A^X$.

$$\mu_w(x) = \begin{cases} 0' & \text{if } x = x_0, \\ 0 & \text{if } x = x_1, \\ 1 & \text{if } x = w, \\ 0 & \text{if } x \in W - \{w\}, \\ 1 & \text{if } w \in x \in E, \\ 0 & \text{if } w \notin x \in E, \\ x(w) & \text{if } x \in \{0,1\}^W. \end{cases} \tag{6.10}$$

We define two further functions, μ and λ.

$$\lambda(x) = \begin{cases} 0 & \text{if } x \in \{x_0, x_1\}, \\ 1 & \text{otherwise.} \end{cases}$$

$$\mu(x) = \begin{cases} 1 & \text{if } x \in \{0,1\}^W, \\ 0 & \text{otherwise.} \end{cases} \tag{6.11}$$

Next we define

$$\mathbf{D} \text{ is the subalgebra of } \mathbf{A}^X \text{ generated by}$$

$$\{\mu_w : w \in W\} \cup \{\mu, \lambda\} \cup \{\hat{a} : a \in A\}. \tag{6.12}$$

We shall write $\bar{a}'$ for the sequence $\langle \hat{a}_0, \ldots, \hat{a}_{k-1}\rangle$ of functions corresponding to the elements of $\mathbf{A}$. We put $e^{\mathbf{D}}(f) = e^{(X)}|_D(f)$ $(= t^{\mathbf{D}}(f, \bar{a}'))$ and $e_i^{\mathbf{D}}(f) = e_i^{(X)}|_D(f)$ $(= t_i^{\mathbf{D}}(f, \bar{a}'))$. (See (6.8).) Thus we have

$$e^{\mathbf{D}},\ e_i^{\mathbf{D}} \in \mathrm{E}(\mathbf{D}),\ e^{\mathbf{D}}(D) = D(M),\ \text{ and } e_i^{\mathbf{D}}(D) = D(N_i). \tag{6.13}$$

We shall need these facts:

$$\begin{gathered}\text{Let } f \in D \text{ and } x, y \in X. \text{ Then } \langle f(x), f(y)\rangle \in \beta;\\ \text{moreover if } x \neq x_0 \neq y \text{ and } f(x) \neq f(y)\\ \text{then } \{f(x), f(y)\} \in M_{\mathbf{A}}(\alpha, \beta).\end{gathered} \tag{6.14}$$

The truth of the first claim is obvious since all of the generating functions for **D** have the property. To prove the second claim, suppose that $x \neq x_0 \neq y$ and $f(x) \neq f(y)$. There exist an $n+2$-ary polynomial operation $H(u, v, \bar{z})$ of **A**, for some n, and elements $w_0, \ldots, w_{n-1} \in W$ such that

$$f = H^{(X)}(\mu, \lambda, \mu_{w_0}, \ldots, \mu_{w_{n-1}}).$$

Since $\langle \mu(x), \lambda(x), \bar{\mu}_w(x)\rangle$ and $\langle \mu(y), \lambda(y), \bar{\mu}_w(y)\rangle$ are sequences of 0's and 1's, and $\mathbf{A}|_N$ is polynomially equivalent to a Boolean algebra, there exist polynomial functions $r, s, r_0, \ldots, r_{n-1}$ of **A** such that these sequences are, respectively, $\langle r(0), s(0), \bar{r}(0)\rangle$, and $\langle r(1), s(1), \bar{r}(1)\rangle$. Then letting $h(u) = H(r(u), s(u), \bar{r}(u))$, we have that $\langle f(x), f(y)\rangle = \langle h(0), h(1)\rangle$. Now our initial assumption about 0 and 1 implies that $\langle h(0), h(1)\rangle \in \beta - \alpha$. Thus by Theorem 0.25 (iii), $\{h(0), h(1)\} \in M_{\mathbf{A}}(\alpha, \beta)$. This finishes our proof of (6.14).

Now since $\mathbf{A}|_N$ has the operations of a Boolean algebra, the set $D(N)$ is closed under the Boolean operations. This set, and the Boolean operations and the order relation on it, are definable in the algebra **D**. We introduce a definable subset L of D.

$$\begin{gathered}f \in L \overset{\text{def}}{\longleftrightarrow} f \in D(M)\ \wedge\ e_0^{\mathbf{D}}(f) \leq \lambda\ \wedge\ \hat{0} \neq e_0^{\mathbf{D}}(f) \wedge\\ \textstyle\bigwedge_{i<r}(e_i^{\mathbf{D}}(f) \neq f)\ \wedge\ \bigwedge_{i<r}(e_i^{\mathbf{D}} e_0^{\mathbf{D}}(f) = e_i^{\mathbf{D}}(f)).\end{gathered} \tag{6.15}$$

The crux of this proof lies in the next claim.

Claim: Every function μ_w, $w \in W$, belongs to L. If $f \in L$ then there exists $w \in W$ such that $e_0^{\mathbf{D}}(f) \leq e_0^{\mathbf{D}}(\mu_w)$ on the set $X - \{0, 1\}^W$.

It should be obvious after a little reflection that $\mu_w \in L$. Notice that $e_i^{\mathbf{D}}(\mu_w) = \mu_w$ cannot hold for any $i < r$ because the range of $e_i^{\mathbf{A}}$ is a two-element set, while the range of μ_w is the set $\{0', 0, 1\}$.

To prove the harder part of the claim, let $f \in L$. Then put

$$T = f(X - \{x_0\}). \tag{6.16}$$

We assert that $T \in M_{\mathbf{A}}(\alpha, \beta)$. To prove this assertion, first recall that by (6.14) we have that every two distinct elements u, v of T constitute a set $N_i \in M_{\mathbf{A}}(\alpha, \beta)$. Then where $e_i(A) = N_i$, since $e_i^{\mathbf{D}} e_0^{\mathbf{D}}(f) = e_i^{\mathbf{D}}(f)$, it follows that $e_0(u) \neq e_0(v)$. Thus T has no more elements than does $e_0(A) = N$; i.e., it has at most two elements. To see that T has more than one element, suppose otherwise; say $T = \{u\}$. Then $f(x_0) \equiv f(x_1) = u \pmod{\alpha}$ (since all the generating functions of $\mathbf{D}$ have this property) and hence $f(X) \subseteq u/\alpha$. But this implies that $e_0^{\mathbf{D}}(f)$ is constant, which is ruled out by (6.15). Now (6.14) implies that $T \in M_{\mathbf{A}}(\alpha, \beta)$, as we asserted.

From the proof of the assertion it follows that we can write

$$\begin{gathered} T = f(X - \{x_0\}) = \{0^\star, 1^\star\} = e_i(A)\,, \\ \text{where } f(x_1) = 0^\star\,, \\ e_0(0^\star) = 0 \text{ and } e_0(1^\star) = 1. \end{gathered} \tag{6.17}$$

(Note that $e_0(0^\star) = 0$ because $e_0^{\mathbf{D}}(f) \leq \lambda$.) We have

$$f(x_0) = 0^\circ \text{ where } \langle 0^\star, 0^\circ \rangle \in \alpha - 0_A \tag{6.18}$$

since $e_i^{\mathbf{D}}(f) \neq f$.

Now let us write

$$f = H^{(X)}(\mu, \lambda, \mu_{w_0}, \ldots, \mu_{w_{n-1}}) \tag{6.19}$$

where $\{w_0, \ldots, w_{n-1}\} \subseteq W$ and $H \in \mathrm{Pol}_{n+2}\mathbf{A}$. Since $f \in D(M) = e^{\mathbf{D}}(D)$, we can assume that $eH = H$, i.e., that the range of H is included in M. Observe that by (6.10), (6.11), (6.17) and (6.18) we have $H(0, 0, 0', \ldots, 0') = f(x_0) = 0^\circ$ and $H(0, 0, 0, \ldots, 0) = f(x_1) = 0^\star$. The set $H(\{0, 0'\}^{n+2}) \subseteq M$ is contained in one $\langle 0_A, \alpha \rangle$-trace. Thus by Lemma 0.40 (since $\mathrm{typ}(0_A, \alpha) = \mathbf{1}$), H restricted to $\{0, 0'\}^{n+2}$ depends on only one variable; obviously that variable is different from the first and the second variables.

By permuting the variables in H, we can assume that H depends on just its third variable when all its arguments are restricted to the set $\{0, 0'\}$. We proceed to show that $e_0^{\mathbf{D}}(f) \leq e_0^{\mathbf{D}}(\mu_{w_0})$ on the set $X - \{0, 1\}^W$. So let $x \in X - \{0, 1\}^W$. The desired inequality is already known if $x \in \{x_0, x_1\}$; so we assume further that $x \in W \cup E$ and that $\mu_{w_0}(x) = 0$ and $f(x) = 1^\star$, in order to obtain a contradiction. (See (6.17).) We also assume that the variables of H are so ordered that $\mu_{w_1}(x) = \cdots = \mu_{w_m}(x) = 1$ while $\mu_{w_{m+1}}(x) = \cdots = \mu_{w_{n-1}}(x) = 0$. We now define an auxiliary polynomial operation,

$$S(y,z) = H(y,z,y,z,\ldots,z,0,\ldots,0)$$

in which z occurs in the places reserved for λ and $\mu_{w_1},\ldots,\mu_{w_m}$.

Notice that since H restricted to $\{0,0'\}$ depends only on its third variable, we have

$$\begin{gathered} S(0',0) = S(0',0') = f(x_0) = 0^\circ, \text{ while} \\ S(0,0) = f(x_1) = 0^\star; \text{ also} \\ S(0,1) = f(x) = 1^\star. \end{gathered} \tag{6.20}$$

Recall that there exists an element $r \in \{0,1\}^W \subseteq X$ such that

$$\begin{gathered} \mu(r) = \lambda(r) = \mu_{w_0}(r) = \cdots = \mu_{w_m}(r) = 1 \text{ and} \\ \mu_{w_{m+1}}(r) = \cdots = \mu_{w_{n-1}}(r) = 0. \end{gathered}$$

Thus we have $S(1,1) = f(r) \in \{0^\star, 1^\star\}$.

Now we can easily rule out the possibility that $S(1,1) = 0^\star$. For if this holds, then the polynomial function $U(z) = S(z,z)$ satisfies $U(0') = 0^\circ$, $U(0) = U(1) = 0^\star$. Since $\langle 0,0'\rangle \in \alpha - 0_A$ and $0^\circ \neq 0^\star$, Theorem 0.25 (iii) implies that the function U defines a permutation on M, which is, however, contradicted by $U(0) = U(1)$. Therefore we must have $S(1,1) = S(0,1) = 1^\star$. Hence the polynomial function $V(z) = S(z,1)$ is not a permutation on M, and again by Theorem 0.25, we have $S(0,1) = S(0',1)$. Recall that $\alpha \leq Z(\mathbf{A})$ (by Corollary 3.2) and $\langle 0,1\rangle \in \beta$. From this it follows by (6.6) that the equation $S(0,1) = S(0',1)$ implies $S(0,0) = S(0',0)$. But this contradicts (6.20) and so ends our proof of the main claim.

The proof of this lemma can now be quickly concluded. We put

$$L' = \{e_0^{\mathbf{D}}(f) \vee \mu : f \in L\}.$$

For $w \in W$ we put $\gamma_w = e_0^{\mathbf{D}}(\mu_w) \vee \mu$, and we define

$$\Gamma = \{\gamma_w : w \in W\}.$$

Obviously the set $L' \subseteq D(N)$ is definable and by the claim, Γ is precisely the set of maximal members of L'. Let $\xi \in D$ be the Boolean complement in $D(N)$ of the element μ. Now it is easy to check that for $w \neq w' \in W$ we have $\{w,w'\} \in E$ iff $\mathbf{D} \models \mathrm{Edg}(\gamma_w, \gamma_{w'})$ where $\mathrm{Edg}(y,z)$ is the formula

$$y \wedge z \wedge \xi \neq \hat{0} \text{ in } D(N).$$

This observation concludes the proof that **W** is interpretable into **D**. It should be clear that the interpretation we have constructed is independent of the choice of **W**, so that we have actually interpreted the class of all graphs into $\boldsymbol{P}_s(\mathbf{A})$. □

To finish our proof of Theorem 6.1, it remains to rule out the existence of a finite algebra **A** in $\mathcal{V}$ having a three-element interval $I[0_A, \beta] = \{0_A, \alpha, \beta\}$ of type $(\mathbf{1}, \mathbf{3})$ or $(\mathbf{2}, \mathbf{3})$ in **Con A**, and having $N = \{0, 1\}$ in $M_{\mathbf{A}}(\alpha, \beta)$ and $f \in \mathrm{Pol}_1 \mathbf{A}$ satisfying $\langle f(0), f(1)\rangle \in \alpha - 0_A$. Here is the last lemma we need.

LEMMA 6.5 *Let* **A** *be a finite algebra, and suppose that* **A** *has elements* $a_1 \neq a_0 \neq a_2$ *and* $b_1 \neq b_0 \neq b_2$ *and polynomial functions* p, e_1, e_2, q *satisfying these conditions.*

(i) $p(b_i) = a_i$ *for* $i = 0, 1, 2$.

(ii) $e_1, e_2 \in \mathrm{E}(\mathbf{A})$ *and* $e_1(A) = \{b_0, b_1\}$ *and* $e_2(A) = \{b_0, b_2\}$.

(iii) $\mathbf{A}|_{\{b_0, b_1\}}$ *and* $\mathbf{A}|_{\{b_0, b_2\}}$ *are minimal algebras of type* **3**.

(iv) *For all* $h \in \mathrm{Pol}_1 \mathbf{A}$, *the function* $e_2 h e_1$ *is constant.*

(v) $q(\{a_0, a_1\}) = \{a_0, a_2\}$.

Then $\boldsymbol{P}_s(\mathbf{A})$ *is an unstructured class.*

PROOF. We shall interpret $\mathcal{BP}_1$ (Definition 0.46) into $\boldsymbol{P}_s(\mathbf{A})$. Let $\mathbf{B} = \langle B_1, B_2, \leq\rangle$ be a member of $\mathcal{BP}_1$ with $B_1 \subseteq \mathcal{P}(X)$. For $i \in \{1, 2\}$, put $N_i = \{b_0, b_i\}$. Then define $\mathbf{D} \subseteq \mathbf{A}^X$ to be the subalgebra of $\mathbf{A}^X$ generated by

$$N_1[B_1] \cup N_2[B_2] \cup \{\hat{a} : a \in A\}$$

where, as usual, $U[B_i]$ denotes the set of functions $f \in A^X$ such that $f^{-1}(u) \in B_i$ for all $u \in U$, and $\hat{a}$ denotes the constant function in A^X correlated with the element $a \in A$.

We wish to build an interpretation of **B** into **D**. It is actually very easy to do so. We shall use $p^{\mathbf{D}}, e_1^{\mathbf{D}}, e_2^{\mathbf{D}}, q^{\mathbf{D}}$ to denote the functions on D which act coordinatewise like p, e_1, e_2 and q. Since **D** is a diagonal subalgebra of $\mathbf{A}^X$, these functions map D into D and are definable in **D**.

Clearly $D \subseteq A[B_1]$, implying

$$e_1^{\mathbf{D}}(D) = D(N_1) = N_1[B_1]. \tag{6.21}$$

We claim that

$$e_2^{\mathbf{D}}(D) = D(N_2) = N_2[B_2]. \tag{6.22}$$

The proof of (6.22) relies on condition (iv). It is clear that $e_2^{\mathbf{D}}(D) = D(N_2)$. At issue is the assertion that $D(N_2) = N_2[B_2]$. It is also clear that $N_2[B_2] \subseteq D(N_2)$. To see that $D(N_2) \subseteq N_2[B_2]$, let $f \in D(N_2)$. By definition of $\mathbf{D}$, there exist $\bar{h} \in (N_1[B_1])^m$ and $\bar{k} \in (N_2[B_2])^n$ for some positive m and n, and a polynomial operation $H(\bar{x}, \bar{y})$ of $\mathbf{A}$, so that

$$f = H^{(X)}(\bar{h}, \bar{k}).$$

We are assuming that $f(X) \subseteq N_2 = \{b_0, b_2\}$, and thus we can also assume that $H = e_2 H$ (or replace H by $e_2 H$).

Now choosing any $u, v \in X$ we can produce—by the same method used near the end of the proof of Lemma 6.2, in view of condition (iii)—a polynomial function h of $\mathbf{A}$ such that

$$h(b_0) = H(\bar{h}(u), \bar{k}(u)) \text{ and } h(b_1) = H(\bar{h}(v), \bar{k}(u)).$$

Because $H(\bar{h}(u), \bar{k}(u)), H(\bar{h}(v), \bar{k}(u)) \in N_2$ it follows from (iv) that

$$H(\bar{h}(u), \bar{k}(u)) = H(\bar{h}(v), \bar{k}(u)).$$

So now choosing a fixed $v_0 \in X$, we have that

$$f(u) = H(\bar{h}(v_0), \bar{k}(u))$$

for all $u \in X$. This implies that there is a term t and constant functions $\epsilon_0, \ldots, \epsilon_{r-1}$ for some r so that

$$f = t^{\mathbf{D}}(\bar{\epsilon}, \bar{k}).$$

The above equation clearly implies that $f \in N_2[B_2]$, and this ends our proof of (6.22).

In view of (iii) and (6.21), the algebra $\mathbf{N}_1 = \mathbf{A}|_{N_1}$ has the operations of a Boolean algebra in which $b_0 < b_1$; and $D(N_1)$ is a Boolean subuniverse of $\mathbf{N}_1^X$ inheriting an order $\leq$ such that $\langle D(N_1), \leq \rangle \cong \langle B_1, \leq \rangle$. The set $D(N_1)$ and the order on it are definable in $\mathbf{D}$. The natural isomorphism of $\langle D(N_1), \leq \rangle$ with $\langle B_1, \leq \rangle$ maps $N_1[B_2]$ onto B_2. All that remains is to show that $N_1[B_2]$, which is the set of functions $f \in D(N_1)$ such that $f^{-1}(b_1) \in B_2$, is definable in $\mathbf{D}$. In fact, for $f \in D(N_1)$, we have

$$f \in N_1[B_2] \longleftrightarrow (\exists g \in D(N_2))[q^{\mathbf{D}} p^{\mathbf{D}}(f) = p^{\mathbf{D}}(g)].$$

It is quite easy to verify the truth of this assertion, using (6.22) and (i) and (v) (and the assumption that $a_1 \neq a_0 \neq a_2$). □

Proof of Theorem 6.1. Suppose that the theorem is not true. We shall reach a contradiction. There exists a finite algebra $\mathbf{F} \in \mathcal{V}$ with a three-element interval $I[\chi_0, \chi_2]$ of type $(\mathbf{1},\mathbf{3})$ or $(\mathbf{2},\mathbf{3})$ in $\mathbf{Con\,F}$. We can assume that $\chi_0 = 0_F$ (or take the quotient algebra $\mathbf{F}/\chi_0$). Thus we have a three-element interval $I[0_F, \beta]$ where $0_F \prec \alpha \prec \beta$, $\mathrm{typ}(\alpha,\beta) = \mathbf{3}$ and α is an Abelian congruence (i.e., $\mathrm{typ}(0_F, \alpha) \in \{\mathbf{1},\mathbf{2}\}$).

Let $N = \{0,1\}$ be a fixed member of $M_{\mathbf{F}}(\alpha,\beta)$. By either Lemma 6.2 or Lemma 6.4, there exists $f \in \mathrm{Pol}_1\mathbf{F}$ such that $\langle f(0), f(1)\rangle \in \alpha - 0_F$. By Theorem 0.25 (iv) we can assume that $f(F) \subseteq M$ for some $M \in M_{\mathbf{F}}(0_F, \alpha)$. Let us put $0^\circ = f(0)$, $1^\circ = f(1)$. Our goal is to construct a finite algebra $\mathbf{A} \in \mathcal{V}$ for which it will follow from Lemma 6.5 that $\boldsymbol{P}_s(\mathbf{A})$ is unstructured. $\mathbf{A}$ will be a quotient algebra of $\mathbf{F}^2$.

We define two congruences on $\mathbf{F}^2$.

$$\begin{aligned} \Delta &= \mathrm{Cg}^{\mathbf{F}^2}(\{\langle (x,x),(y,y)\rangle : (x,y) \in \alpha\}), \\ \alpha \times \alpha &= \{\langle (x,y),(u,v)\rangle : (x,u) \in \alpha \ \wedge\ (y,v) \in \alpha\}. \end{aligned} \tag{6.23}$$

Notice that $\Delta \subseteq \alpha \times \alpha$ and $\alpha \times \alpha$ is an Abelian congruence. Thus $\alpha \times \alpha$ is solvable over Δ (by Theorem 0.11). Then by Corollary 3.2, $(\alpha \times \alpha)/\Delta$ is contained in the center of $\mathbf{F}^2/\Delta$. This is equivalent to

$$\alpha \times \alpha \subseteq (\Delta : 1_{F^2}). \tag{6.24}$$

By the same token, $\alpha \subseteq Z(\mathbf{F})$ and $\alpha \times \alpha \subseteq Z(\mathbf{F}^2)$.

The next assertion is a consequence of the fact that $\alpha \subseteq Z(\mathbf{F})$. We leave its proof to the reader.

$$(x,x)/\Delta = \{(y,y) : (x,y) \in \alpha\} \text{ for all } x \in F. \tag{6.25}$$

The argument now divides into the consideration of two cases.

CASE (I): $\mathbf{F}$ has an $m+1$-ary polynomial operation $k(x,\bar{y})$ (for some m) such that there exist $\bar{r}, \bar{s} \in F^m$ satisfying : $k(F^{m+1}) \subseteq M$ and $k(1^\circ, \bar{r}) \neq k(0^\circ, \bar{r}) = k(1^\circ, \bar{s}) = 0^\circ$.

Given such $k, \bar{r}, \bar{s}$ we define $k_{\bar{r}}(x) = k(x, \bar{r})$. Then by Theorem 0.25, $k_{\bar{r}}$ restricted to M is a permutation of M since $(0^\circ, 1^\circ) \in \alpha - 0_M$. Thus there exists a polynomial function π (a power of $k_{\bar{r}}$) such that

$$\pi|_M = (k_{\bar{r}}|_M)^{-1}. \tag{6.26}$$

Let $a = k(1^\circ, \bar{r})$, $b = k(0^\circ, \bar{s})$. For $\bar{x}, \bar{y} \in F^m$, denote by $(\bar{x}, \bar{y})$ the m-tuple $\langle (x_0, y_0), \ldots, (x_{m-1}, y_{m-1})\rangle \in (F^2)^m$. Now we have

$$k^{\mathbf{F}^2}((1^\circ,1^\circ),(\bar{r},\bar{r})) \equiv_\Delta k^{\mathbf{F}^2}((0^\circ,0^\circ),(\bar{r},\bar{r})) = k^{\mathbf{F}^2}((1^\circ,1^\circ),(\bar{s},\bar{s})).$$

By (6.24) it follows that

$$k^{\mathbf{F}^2}((1^\circ,0^\circ),(\bar{r},\bar{r})) \equiv_\Delta k^{\mathbf{F}^2}((1^\circ,0^\circ),\bar{s},\bar{s})),$$

i.e., that

$$(a,0^\circ) \equiv_\Delta (0^\circ,b). \tag{6.27}$$

Now we introduce another unary polynomial function of $\mathbf{F}^2$. Let

$$g(x,y) = \pi^{\mathbf{F}^2} k^{\mathbf{F}^2}((x,y),(\bar{r},\bar{s})) = (\pi k(x,\bar{r}), \pi k(y,\bar{s})).$$

Notice that $\pi(0^\circ) = 0^\circ$ and $\pi(a) = 1^\circ$ (since $k(1^\circ,\bar{r}) = a$). Using (6.27) we deduce that

$$\begin{gathered} g(0^\circ,1^\circ) = (0^\circ,0^\circ), \text{ and} \\ g(0^\circ,0^\circ) = \pi^{\mathbf{F}^2}(0^\circ,b) \equiv_\Delta \pi^{\mathbf{F}^2}(a,0^\circ) = (1^\circ,0^\circ). \end{gathered} \tag{6.28}$$

Now we take $\mathbf{A} = \mathbf{F}^2/\Delta$. Elements $a_0,\ldots,b_2$ of $\mathbf{A}$ are defined as follows. We put $a_0 = a_0'/\Delta,\ldots,b_2 = b_2'/\Delta$ where

$$\begin{array}{ll} a_0' = (0^\circ,0^\circ), & b_0' = (0,0) \\ a_1' = (0^\circ,1^\circ), & b_1' = (0,1) \\ a_2' = (1^\circ,0^\circ), & b_2' = (1,0)\,. \end{array} \tag{6.29}$$

It follows from (6.25) that $a_1 \neq a_0 \neq a_2$ and $b_1 \neq b_0 \neq b_2$. Moreover, $f^{\mathbf{F}^2}(b_i') = a_i'$ for $i = 0,1,2$ since $f(0) = 0^\circ$ and $f(1) = 1^\circ$. We define polynomial functions p, e_1, e_2, q of $\mathbf{A}$ as follows. Let $e \in \mathrm{E}(\mathbf{F})$ satisfy $e(F) = N = \{0,1\}$; and let $m(x,y)$ be a binary polynomial operation of $\mathbf{F}$ such that $m|_N$ is the meet operation on N.

$$\begin{aligned} p((x,y)/\Delta) &= (f(x),f(y))/\Delta \\ e_1((x,y)/\Delta) &= m^{\mathbf{F}^2}(b_1',(e(x),e(y)))/\Delta \\ e_2((x,y)/\Delta) &= m^{\mathbf{F}^2}(b_2',(e(x),e(y)))/\Delta \\ q((x,y)/\Delta) &= g(x,y)/\Delta. \end{aligned} \tag{6.30}$$

It can immediately be verified that conditions (i), (ii), and (v) of Lemma 6.5 hold, using (6.28), the fact that $f(0) = 0^\circ$ and $f(1) = 1^\circ$, and the definitions of e, e_1, e_2 etc. To verify (iii), observe that $\mathbf{F}^2|_{\{b_0',b_1'\}}$ has all possible operations on the two-element set, because $\mathbf{F}|_N$ has all possible operations on the set $\{0,1\}$. By (6.25), $\langle b_0',b_1'\rangle \notin \Delta$, and from this it follows that $\mathbf{A}|_{\{b_0,b_1\}}$ is a two-element minimal algebra of type **3**. For $\mathbf{A}|_{\{b_0,b_2\}}$, the proof is the same.

To verify condition (iv) of Lemma 6.5, let $h \in \mathrm{Pol}_1\mathbf{A}$. We must show that e_2he_1 is constant. Equivalently, we must show that $e_2h(b_0) = e_2h(b_1)$. There is a polynomial function $h' \in \mathrm{Pol}_1\mathbf{F}^2$ such that $h(x/\Delta) = h'(x)/\Delta$ for all $x \in F^2$. Then for $i \in \{0,1\}$, we have $e_2h(b_i) = c_i'/\Delta$ where

$$c_i' = m^{\mathbf{F}^2}(b_2', e^{\mathbf{F}^2}h'(b_i')).$$

Since b_0' and b_1' have the same left components, so do c_0' and c_1'. Clearly, c_0' and c_1' have the same right components, since they are members of the Boolean algebra $(\mathbf{F}|_N)^2$ and are both $\le (1,0)$. Hence $c_0' = c_1'$ and $e_2h(b_0) = e_2h(b_1)$.

This ends our proof in Case (I). We have constructed an algebra $\mathbf{A} \in \mathcal{V}$ for which Lemma 6.5 implies that $\boldsymbol{P}_s(\mathbf{A})$ is unstructured.

CASE (II): This is the negation of Case (I).

In this case, we introduce another congruence Γ on $\mathbf{F}^2$.

$$\Gamma = \mathrm{Cg}^{\mathbf{F}^2}((0^\circ, 1^\circ), (1^\circ, 0^\circ)).$$

We take $\mathbf{A} = \mathbf{F}^2/\Gamma$ and define $a_i = a_i'/\Gamma$ and $b_i = b_i'/\Gamma$ where $a_0', \dots, b_2'$ are the elements defined in (6.29). Polynomial functions p, e_1, e_2, q of $\mathbf{A}$ are defined just as in (6.30) with only one difference; namely, we put $q(x) = x$ for $x \in A$.

Since $\Gamma \subseteq \alpha \times \alpha$, then b_0, b_1, b_2 are three distinct elements. We now have $a_1 = a_2$ since $\Gamma = \mathrm{Cg}^{\mathbf{F}^2}(a_1', a_2')$. All the conditions of Lemma 6.5 can be verified in the same way as was done in Case (I), with one exception. A special argument is needed to show that $a_0 \ne a_1$. This will follow from

$$M^2 \cap ((0^\circ, 0^\circ)/\Gamma) = \{(0^\circ, 0^\circ)\}, \tag{6.31}$$

which we now prove.

Suppose that $(a,b) \in M^2 \cap ((0^\circ, 0^\circ)/\Gamma)$. Now M^2 is the range of an idempotent polynomial function of $\mathbf{F}^2$, and $\Gamma = \mathrm{Cg}^{\mathbf{F}^2}(a_1', a_2')$. Thus $\Gamma \cap (M^2)^2$ is the transitive symmetric closure of the relation R consisting of the pairs $\langle x, y\rangle \in (M^2)^2$ such that $\langle x, y\rangle = \langle h(a_1'), h(a_2')\rangle$ for some $h \in \mathrm{Pol}_1\mathbf{F}^2$ with $h(F^2) \subseteq M^2$. Thus if $(a,b) \ne (0^\circ, 0^\circ)$ then there must exist such an h with say

$$h(a_1') = (0^\circ, 0^\circ) \text{ and } h(a_2') \in M^2 - \{(0^\circ, 0^\circ)\}. \tag{6.32}$$

(The case where $h(a_2') = (0^\circ, 0^\circ)$ is completely symmetric to this one, and can be handled by the same argument.) Let us assume that (6.32) holds. We can write

$$h((x,y)) = (k(x, \bar{r}), k(y, \bar{s}))$$

where k is an $m+1$ -ary polynomial operation of $\mathbf{F}$ (for some m), $\bar{r}$ and $\bar{s}$ belong to F^m, and $k(F^{m+1}) \subseteq M$.

From (6.32) we have

$$k(0^\circ, \bar{r}) = k(1^\circ, \bar{s}) = 0^\circ \text{ and}$$

$$k(1^\circ, \bar{r}) \neq 0^\circ \text{ or } k(0^\circ, \bar{s}) \neq 0^\circ.$$

Since we are not in Case (I), we have that $k(1^\circ, \bar{r}) = 0^\circ$. Thus $k(1^\circ, \bar{s}) = k(1^\circ, \bar{r})$, implying that $k(0^\circ, \bar{s}) = k(0^\circ, \bar{r}) = 0^\circ$ since $\langle 0^\circ, 1^\circ \rangle \in Z(\mathbf{F})$. This is a contradiction, and it proves (6.31). The proof of Theorem 6.1 is finished. □

Summary of Part I

For any locally finite variety $\mathcal{V}$, we defined in Chapter 1 three subvarieties, $\mathcal{V}_1$, $\mathcal{V}_2$ and $\mathcal{V}_3$; and we showed that if $\mathcal{V}$ is structured, then every algebra in $\mathcal{V}$ is a subdirect product of three algebras belonging to these subvarieties. For structured $\mathcal{V}$, we proved in Part I that $\mathcal{V}_3$ is a discriminator variety (Theorem 4.1), and that $\mathcal{V}_1 \vee \mathcal{V}_2$ is an Abelian variety (Theorem 5.4). In Part II, we shall determine the nature of $\mathcal{V}_1$ and $\mathcal{V}_2$, for structured $\mathcal{V}$.

Part II

Structured Abelian varieties

Chapter 7

Strongly solvable varieties

The plan of Part II is as follows. We deal with a locally finite, structured, Abelian variety $\mathcal{V}$. According to Theorem 1.3, we have $\mathcal{V} = \mathcal{V}_1 \vee \mathcal{V}_2$ where $\mathcal{V}_1$ and $\mathcal{V}_2$ are the subvarieties of $\mathcal{V}$ that are defined in Definition 1.1. The first principal result of Part II is achieved in Theorem 9.6: $\mathcal{V}_1$ is strongly Abelian and $\mathcal{V}_2$ is affine. The second principal result is Theorem 11.9: $\mathcal{V}_1$ is equivalent to a structured variety of multi-sorted unary algebras. The third principal result is Theorem 12.19, which characterizes the decidable, locally finite, strongly solvable varieties in terms of an elementary property of the associated variety of multi-sorted unary algebras.

Chapters 7–10 are occupied with the proof of Theorem 9.6. We prove in Chapter 8 that $\mathcal{V}$ satisfies the $(\mathbf{1},\mathbf{2})$ and the $(\mathbf{2},\mathbf{1})$ transfer principles, using three interpretations whose lengthy details are postponed until Chapter 10. Using the transfer principles, we show in Chapter 9 that $\mathcal{V}_2$ is affine and $\mathcal{V}_1$ is locally strongly solvable. In this chapter, we prove that a locally strongly solvable Abelian variety is strongly Abelian. In Valeriote [1986] this was shown using the added assumption that the variety was locally finite. Valeriote along with Bradd Hart managed to remove this added hypothesis, and we present their proof in this chapter. Thus we will be able to conclude in Chapter 9 that $\mathcal{V}_1$ is strongly Abelian.

As was noted in Theorem 0.15, for a locally finite variety $\mathcal{W}$, the class $\mathcal{S}$ of all locally strongly solvable algebras in $\mathcal{W}$ is a variety. Our purpose in this chapter is to show that if $\mathcal{W}$ is Abelian then $\mathcal{S}$ must be strongly Abelian. In fact we will prove the following theorem.

THEOREM 7.1 *Let* $\mathbf{A}$ *be a locally strongly solvable algebra and suppose that every algebra in* $\mathbf{H}\mathbf{S}(\mathbf{A}^2)$ *is Abelian. Then* $\mathbf{A}$ *is strongly Abelian.*

Before proving this theorem, we state an obvious corollary of it and then prove several lemmas that will be needed in the proof.

COROLLARY 7.2 *Let $\mathcal{W}$ be a locally strongly solvable variety. Then $\mathcal{W}$ is strongly Abelian iff $\mathcal{W}$ is Abelian.*

LEMMA 7.3 *Let* **B** *be an algebra and let $\theta \in \mathrm{Con}\mathbf{B}$ be a strongly Abelian congruence. If there is a polynomial operation $t(x, \bar{y})$ of* **B** *and elements a, b, $\bar{c}$ and $\bar{d}$ such that*

$$t(a, \bar{c}) = t(b, \bar{d}) \text{ and } \langle t(a, \bar{c}), t(b, \bar{c})\rangle \in \theta - 0_B$$

then $\boldsymbol{HS}(\mathbf{B}^2)$ contains a non-Abelian algebra.

PROOF. Let θ, $t(x, \bar{y})$, a, b, $\bar{c}$ and $\bar{d}$ be given as in the statement of this lemma. Throughout this proof we may assume that the algebras in $\boldsymbol{HS}(\mathbf{B})$ are Abelian. Let $t(a, \bar{c}) = t(b, \bar{d}) = 0$, $t(b, \bar{c}) = 0'$ and $t(a, \bar{d}) = 0''$. Since the algebra $\mathbf{B}/\theta$ is Abelian then it follows that 0, $0'$ and $0''$ are all θ related.

We claim that if $p(x)$ is a unary polynomial operation of **B** then $p(0) = p(0')$ if and only if $p(0) = p(0'')$. If not then we would have

$$p(t(a, \bar{c})) = p(t(b, \bar{c}))$$

and

$$p(t(a, \bar{d})) \neq p(t(b, \bar{d}))$$

or vice versa, which contradicts our assumption that **B** is Abelian.

Let **C** be the subalgebra of $\mathbf{B}^2$ generated by the pair $\langle a, b\rangle$ and the set $\{\langle u, u\rangle : u \in B\}$ and let γ be the principal congruence on **C** generated by identifying the pairs $\langle 0, 0'\rangle$ and $\langle 0'', 0\rangle$.

We'll show that $\langle\langle 0, 0\rangle, \langle 0', 0'\rangle\rangle \notin \gamma$; thereby establishing that $\mathbf{C}/\gamma$ is non-Abelian. This is because in **C** we have:

$$\begin{aligned}
t(\langle a, b\rangle, \bar{c}^{\diamond}) &= \langle 0, 0'\rangle \\
t(\langle a, b\rangle, \bar{d}^{\diamond}) &= \langle 0'', 0\rangle \\
t(\langle b, b\rangle, \bar{c}^{\diamond}) &= \langle 0', 0'\rangle \\
t(\langle b, b\rangle, \bar{d}^{\diamond}) &= \langle 0, 0\rangle,
\end{aligned}$$

where $\bar{c}^{\diamond}$ and $\bar{d}^{\diamond}$ denote the k-tuples

$$\langle\langle c_1, c_1\rangle, \ldots, \langle c_k, c_k\rangle\rangle,$$

$$\langle\langle d_1, d_1\rangle, \ldots, \langle d_k, d_k\rangle\rangle$$

in **C**.

Suppose to the contrary that $\langle\langle 0,0\rangle,\langle 0',0'\rangle\rangle \in \gamma$ and choose l minimal, $e_0,\ldots,e_l \in C$, $g_0,\ldots,g_{l-1} \in \mathrm{Pol}_1\mathbf{C}$ such that $\langle 0,0\rangle = e_0$, $\langle 0',0'\rangle = e_l$ and $g_i(\{\langle 0,0'\rangle,\langle 0'',0\rangle\}) = \{e_i,e_{i+1}\}$ for $i<l$. Minimality of l ensures that $e_i \neq e_{i+1}$ for all $i<l$.

Now,

$$g_0(\{\langle 0,0'\rangle,\langle 0'',0\rangle\}) = \{e_0,e_1\}$$

so either

$$g_0(\langle 0,0'\rangle) = e_0 \text{ or } g_0(\langle 0'',0\rangle) = e_0.$$

In either case, since $g_0 \in \mathrm{Pol}_1\mathbf{C}$ there is some $n+1$-ary term $s(x,\bar{y})$ and $\bar{u}$, $\bar{v} \in B^n$ with $\langle u_i,v_i\rangle \in C$ for $i \leq n$ such that $g_0(\langle x_0,x_1\rangle) = \langle s(x_0,\bar{u}),s(x_1,\bar{v})\rangle$, for all $\langle x_0,x_1\rangle \in C$.

At this point our proof breaks into two cases, the first being when we can find a polynomial operation $t(x,\bar{y})$ and elements a, b, $\bar{c}$ and $\bar{d}$ as in the statement of this lemma with $\langle a,b\rangle \in \theta$, and the second when this situation never arises.

In the first case, we assume that $\langle a,b\rangle \in \theta$; then if we have $g_0(\langle 0,0'\rangle) = e_0 = \langle 0,0\rangle$, we get that $s(0,\bar{u}) = 0 = s(0',\bar{v})$ in $\mathbf{B}$. Since we are assuming that $\langle a,b\rangle \in \theta$, then for any pair $\langle e,f\rangle \in C$ we have $\langle e,f\rangle \in \theta$. In particular, $\langle u_i,v_i\rangle \in \theta$ for all $i \leq n$.

By assumption, θ is strongly Abelian, and so we conclude that $s(0,\bar{u}) = s(0',\bar{u})$ and $s(0,\bar{v}) = s(0',\bar{v})$. By an earlier claim it follows that $s(0,\bar{u}) = s(0'',\bar{u})$ and so $g_0(\langle 0'',0\rangle) = \langle 0,0\rangle$. But then $e_0 = e_1$, a contradiction. If instead we have $g_0(\langle 0'',0\rangle) = e_0$, then using a similar argument we reach the same contradiction.

In the second case, if we have $g_0(\langle 0,0'\rangle) = e_0$, then we have $s(0,\bar{u}) = s(0',\bar{v}) = 0$. Since $\langle 0,0'\rangle \in \theta$, then because we are in case two, it follows that $s(0,\bar{u}) = s(0',\bar{u})$ and $s(0,\bar{v}) = s(0',\bar{v})$. But then $s(0'',\bar{u}) = s(0,\bar{u}) = 0$ and so $g_0(\langle 0'',0\rangle) = \langle 0,0\rangle = e_0$, a contradiction. The remaining case, when $g_0(\langle 0'',0\rangle) = e_0$ is handled similarly.

Since $\mathbf{C}/\theta$ is in $\boldsymbol{HS}(\mathbf{B}^2)$, the lemma is proved. □

Proof of Theorem 7.1. Assume that $\mathbf{A}$ is a locally strongly solvable algebra that is not strongly Abelian. We must show that some algebra in $\boldsymbol{HS}(\mathbf{A}^2)$ is not Abelian. Without loss of generality, we may assume that $\mathbf{A}$ is finitely generated, and so is actually strongly solvable. Thus there is a chain of congruences $\alpha_0 = 0_A \leq \alpha_1 \leq \cdots \leq \alpha_n = 1_A$ in Con $\mathbf{A}$ such that α_{i+1} is strongly Abelian over α_i, for $i<n$.

Choose $j<n$ maximal with $\mathbf{A}/\alpha_j$ not strongly Abelian. Let $\mathbf{B}$ be $\mathbf{A}/\alpha_j$ and let $\theta = \alpha_{j+1}/\alpha_j$ in Con $\mathbf{B}$. Thus $\mathbf{B}$ is not strongly Abelian, but we may assume that it is Abelian. So, there is a term operation $t(x,\bar{y})$ and elements a, b, $\bar{c}$, $\bar{d}$ of $\mathbf{B}$ such that $t(a,\bar{c}) = t(b,\bar{d})$ and $t(a,\bar{c}) \neq t(b,\bar{c})$. Since

$\mathbf{B}/\theta$ is strongly Abelian, then we have $\langle t(a,\bar{c}), t(b,\bar{c})\rangle \in \theta$. By Lemma 7.3, $\boldsymbol{HS}(\mathbf{B}^2)$ contains a non-Abelian algebra and since $\boldsymbol{HS}(\mathbf{B}^2) \subset \boldsymbol{HS}(\mathbf{A}^2)$, we are done. □

A relativized version of Theorem 7.1 exists, and we state it below without proof.

THEOREM 7.4 *Let* $\mathbf{A}$ *be an algebra such that for every* $\mathbf{C} \in \boldsymbol{HS}(\mathbf{A}^2)$, *every locally strongly solvable interval in* **Con C** *is Abelian. Then every locally strongly solvable interval in* **Con A** *is in fact strongly Abelian.*

Chapter 8

More transfer principles

For a finite solvable algebra $\mathbf{A}$, the only possible labels of prime quotients in $\mathbf{Con\,A}$ are $\mathbf{1}$ and $\mathbf{2}$. Tame congruence theory places some general restrictions on how these labels may be distributed, depending on the shape of $\mathbf{Con\,A}$, but without added assumptions there is much freedom. If $\mathcal{V}(\mathbf{A})$ happens to be decidable (structured), then we will show that $\mathbf{Con\,A}$ must satisfy the $(\mathbf{1}, \mathbf{2})$ and the $(\mathbf{2}, \mathbf{1})$ transfer principles as defined in Chapter 5.

The goal of this chapter is to show that a locally finite structured Abelian variety must satisfy both the $(\mathbf{1}, \mathbf{2})$ transfer principle and its dual, the $(\mathbf{2}, \mathbf{1})$ transfer principle. To prove this, we need several technical lemmas dealing with the interaction between type $\mathbf{1}$ and type $\mathbf{2}$ minimal sets.

LEMMA 8.1 *Let $\mathbf{A}$ be a finite algebra and let $0_A = \alpha_0 \prec \alpha_1 \prec \cdots \prec \alpha_n \prec \beta$ be a chain of congruences on $\mathbf{A}$ such that $\text{typ}(\alpha_i, \alpha_{i+1}) \in \{\mathbf{1}, \mathbf{4}, \mathbf{5}\}$ for $i < n$, and $\text{typ}(\alpha_n, \beta) = \mathbf{2}$. Then for all $\langle \alpha_n, \beta\rangle$-traces N we have $\alpha_n|_N = 0_N$.*

PROOF. We will prove this for the case when $n = 1$, and remark that the general case is handled in a similar fashion. Suppose N is an $\langle \alpha_1, \beta\rangle$-trace and contains elements a, b with $\langle a, b\rangle \in \alpha_1 - 0_A$. Choose $U \in M_{\mathbf{A}}(\alpha_1, \beta)$ and $e \in \mathrm{E}(\mathbf{A})$ such that $e(A) = U$ and $N \subseteq U$. Since $\langle a, b\rangle \in \alpha_1 - 0_A$ and a, $b \in U$ we can connect a to b via a series of overlapping $\langle 0_A, \alpha_1\rangle$-traces contained entirely in $(a/\alpha_1) \cap N$. Thus we may assume that the elements a, b are actually contained in a $\langle 0_A, \alpha_1\rangle$-trace $M \subseteq N$. Since $e = e^2$ then $e(M) = M$ and so there is some $V \in M_{\mathbf{A}}(0_A, \alpha_1)$ such that $M \subseteq V \subseteq U$. Choose $e' \in \mathrm{E}(\mathbf{A})$ with $e'(A) = V$.

Let $d(x, y, z) \in \mathrm{Pol}_3\mathbf{A}$ be an operation such that $d(N^3) = N$ and $d|_N$ is Maltsev (see Lemma 0.38). Let $d'(x, y, z) = e'd(x, y, z)$. Since $d'(a, a, a) = e'(a) = a$, it follows that $d'(M, M, M) \subseteq M$ and $d'|_M$ is a Maltsev operation on M. But this is impossible, since $\text{typ}(0_A, \alpha_1) \in \{\mathbf{1}, \mathbf{4}, \mathbf{5}\}$. Thus $\alpha_1|_N = 0_N$. □

For the remainder of this chapter, let $\mathbf{A}$ be a finite algebra with congruences α and β satisfying $0_A \prec \alpha \prec \beta$.

LEMMA 8.2 *Suppose that $\mathbf{A}$ is solvable, $\text{typ}(0_A, \alpha) = \mathbf{1}$ and $\text{typ}(\alpha, \beta) = \mathbf{2}$. Let M be a $\langle 0_A, \alpha\rangle$-trace and N an $\langle \alpha, \beta\rangle$-trace, and suppose there exists some $f \in \text{Pol}_1\mathbf{A}$ such that $f(N) \subseteq M$ and f is nonconstant on N. Then $\boldsymbol{HS}(\mathbf{A}^2)$ contains a non-Abelian algebra.*

PROOF. We may assume that $\mathbf{A}$ is Abelian. Since N is a type $\mathbf{2}$ trace, then by Lemma 8.1, $\alpha|_N = 0_N$ and so the induced algebra $\mathbf{A}|_N$ is polynomially equivalent to a vector space over some finite field. Choose $0, 1 \in N$ and $x - y \in \text{Pol}_2\mathbf{A}$ such that $\langle f(0), f(1)\rangle \in \alpha - 0_A$ and $x - y$ acts as vector space subtraction for $\mathbf{A}|_N$ with 0 as the additive identity.

Setting $g(x, y) = f(x - y) \in \text{Pol}_2\mathbf{A}$, we have

$$f(0) = g(0, 0) = g(1, 1)$$

and

$$g(0, 0) = f(0)\,(\alpha - 0_A)\,f(1) = g(1, 0).$$

Thus by Lemma 7.3 of the previous chapter, we conclude that $\boldsymbol{HS}(\mathbf{A}^2)$ contains a non-Abelian algebra. □

For any congruence β on $\mathbf{A}$ and subset S of A, by S^β we shall denote the set $\{y : \langle x, y\rangle \in \beta \text{ for some } x \in S\}$.

LEMMA 8.3 *Let $N \subseteq A$ be an $\langle \alpha, \beta\rangle$-trace and choose $0, 1 \in N$ with $\langle 0, 1\rangle \in \beta - \alpha$. Let $U \in M_{\mathbf{A}}(0_A, \alpha)$ and suppose $M \subseteq U$ is a $\langle 0_A, \alpha\rangle$-trace such that $M = M^\beta \cap U$. If $\alpha \subseteq \text{Cg}^{\mathbf{A}}(0, 1)$, then for some $f \in \text{Pol}_1\mathbf{A}$ we have $f(N) \subseteq M$ and f is nonconstant on N.*

PROOF. Choose $a, a' \in M$ distinct. Then $\langle a, a'\rangle \in \alpha$ and so $\langle a, a'\rangle \in \text{Cg}^{\mathbf{A}}(0, 1)$ by assumption. Thus for some $k \in \omega$ we can find $a_0, a_1, \ldots, a_k \in U$ distinct and $g_0, g_1, \ldots, g_{k-1} \in \text{Pol}_1\mathbf{A}$ with $g_i(A) \subseteq U$, $g_i(\{0, 1\}) = \{a_i, a_{i+1}\}$ for $i < k$ and $a_0 = a$, $a_k = a'$.

So $g_0(0) = a_0$ and $g_0(1) = a_1$ or vice versa. In any case we have $\langle a, a_1\rangle \in \beta$ since $\langle 0, 1\rangle \in \beta$, and $a, a_1 \in U$. Thus by our assumption that $M = M^\beta \cap U$, we have $\langle a, a_1\rangle \in \alpha$ and $g_0(N) \subseteq M$, since $a \in M$ and N is contained in a single β class. An inductive argument now shows that $g_i(N) \subseteq M$ for all $i \leq k$; clearly, some one of the functions g_i must be non-constant on N since $a \neq a'$. □

The proof of the next lemma is postponed until Chapter 10, a chapter devoted exclusively to the construction of several elaborate interpretations.

LEMMA 8.4 *Suppose that* $\operatorname{typ}(0_A, \alpha) = \mathbf{1}$, $\operatorname{typ}(\alpha, \beta) = \mathbf{2}$ *and* $\boldsymbol{V}(\mathbf{A})$ *is Abelian. If there is some* $U \in M_{\mathbf{A}}(0_A, \alpha)$ *and a* $\langle 0_A, \alpha\rangle$*-trace* $M \subseteq U$ *with* $M \neq M^\beta \cap U$ *then the class* $\boldsymbol{P}_s(\mathbf{A})$ *is unstructured.*

LEMMA 8.5 *Suppose that* $\operatorname{typ}(0_A, \alpha) = \mathbf{1}$, $\operatorname{typ}(\alpha, \beta) = \mathbf{2}$ *and* $\boldsymbol{V}(\mathbf{A})$ *is Abelian. If for some* $\langle\alpha, \beta\rangle$*-trace* N *and* $0, 1 \in N$ *with* $\langle 0, 1\rangle \in \beta - \alpha$ *we have* $\alpha \subseteq \mathrm{Cg}^{\mathbf{A}}(0, 1)$ *then* $\boldsymbol{P}_s(\mathbf{A})$ *is unstructured.*

PROOF. Suppose $\boldsymbol{P}_s(\mathbf{A})$ is structured for some $\mathbf{A}$ satisfying the hypotheses. Choose $U \in M_{\mathbf{A}}(0_A, \alpha)$. By Lemma 8.4 we may assume $M = M^\beta \cap U$ for all $\langle 0_A, \alpha\rangle$-traces $M \subseteq U$. Let $M \subseteq U$ be any $\langle 0_A, \alpha\rangle$-trace. Lemma 8.3 provides us with an $f \in \mathrm{Pol}_1 \mathbf{A}$ such that $f(N) \subseteq M$ and f is nonconstant on N, since $\alpha \subseteq \mathrm{Cg}^{\mathbf{A}}(0, 1)$. Now Lemma 8.2 tells us that $\boldsymbol{HS}(\mathbf{A}^2)$ contains a non-Abelian algebra, contradicting the fact that $\boldsymbol{V}(\mathbf{A})$ is Abelian. □

It is an easy inference from the previous lemma to prove that any locally finite, structured, Abelian variety satisfies the $(\mathbf{1}, \mathbf{2})$ transfer principle, as will be noted at the end of this chapter. Our proof for the $(\mathbf{2}, \mathbf{1})$ transfer principle roughly parallels the previous one. Again, let $\mathbf{A}$ be a finite algebra with $0_A \prec \alpha \prec \beta$ in $\mathbf{Con}\,\mathbf{A}$. As with Lemma 8.4, the lengthy proof of the following lemma can be found in Chapter 10.

LEMMA 8.6 *Suppose that* $\operatorname{typ}(0_A, \alpha) = \mathbf{2}$, $\operatorname{typ}(\alpha, \beta) = \mathbf{1}$ *and* $\boldsymbol{V}(\mathbf{A})$ *is Abelian. Let* N *be an* $\langle\alpha,\beta\rangle$*-trace and* M *a* $\langle 0_A, \alpha\rangle$*-trace, and suppose there exists some* $f \in \mathrm{Pol}_1 \mathbf{A}$ *such that* $f(N) \subseteq M$ *and* f *is nonconstant on* N*. Then the class* $\boldsymbol{P}_s(\mathbf{A})$ *is unstructured.*

LEMMA 8.7 *Suppose that* $\operatorname{typ}(0_A,\alpha) = \mathbf{2}$, $\operatorname{typ}(\alpha,\beta) = \mathbf{1}$ *and that* $V \in M_{\mathbf{A}}(0_A,\alpha)$*. Then for all* $\langle 0_A,\alpha\rangle$*-traces* $M \subseteq V$ *we have* $M = M^\beta \cap V$*.*

PROOF. Choose $e \in \mathrm{E}(\mathbf{A})$ with $e(A) = V$. Suppose $M \subseteq V$ is a $\langle 0_A, \alpha\rangle$-trace with M properly contained in $M^\beta \cap V$. Choose $u \in M$ and $v \in V$ with $\langle u, v\rangle \in \beta - \alpha$. Since we can connect u to v modulo α with $\langle\alpha,\beta\rangle$-traces in $\mathbf{A}$, then using the idempotent polynomial e we can find some $\langle\alpha,\beta\rangle$-trace $N \subseteq V$ with $N \cap M \neq \emptyset$. Without loss of generality $u \in N \cap M$. Since N is an $\langle\alpha,\beta\rangle$-trace we can find $w \in N$ with $\langle u, w\rangle \in \beta - \alpha$. Choose $U \in M_{\mathbf{A}}(\alpha,\beta)$ with $N \subseteq U \subseteq V$ and let $e' \in \mathrm{E}(\mathbf{A})$ satisfy $e'(A) = U$.

Since $\operatorname{typ}(0_A,\alpha)$ is $\mathbf{2}$, then by Lemma 0.38 there is a polynomial operation $d(x,y,z)$ of $\mathbf{A}$ such that the restriction of d to the body of V is Maltsev. Let $d'(x,y,z) = e'd(x,y,z)$. Since $d'(u,u,u) = e'(u) = u$ and $d'(U^3) \subseteq U$, then $d'(N^3) \subseteq N$. Therefore modulo α, d' acting on N is essentially unary, since $\operatorname{typ}(\alpha,\beta) = \mathbf{1}$. But

$$d'(w,u,u) = e'd(w,u,u) = e'(w) = w \text{ and}$$
$$d'(u,u,w) = e'd(u,u,w) = e'(w) = w,$$

and so modulo α, $d'(x, y, z)$ depends on x and z in N. This contradiction proves the lemma. □

LEMMA 8.8 *Suppose that* $\text{typ}(0_A, \alpha) = \mathbf{2}$, $\text{typ}(\alpha, \beta) = \mathbf{1}$ *and* $\boldsymbol{V}(\mathbf{A})$ *is Abelian. If for some* $\langle \alpha, \beta \rangle$*-trace* N *and* $a, b \in N$ *with* $\langle a, b \rangle \in \beta - \alpha$ *we have* $\alpha \subseteq \text{Cg}^{\mathbf{A}}(a, b)$, *then the class* $\boldsymbol{P}_s(\mathbf{A})$ *is unstructured.*

PROOF. The proof of this lemma uses Lemmas 8.3, 8.6 and 8.7 and is left to the reader. □

Here is the chief result of this chapter.

THEOREM 8.9 *Let* $\mathcal{V}$ *be a locally finite Abelian variety. If* $\mathcal{V}$ *is structured, then* $\mathcal{V}$ *satisfies the* $(\mathbf{1}, \mathbf{2})$ *transfer principle and the* $(\mathbf{2}, \mathbf{1})$ *transfer principle.*

PROOF. Suppose that $\mathcal{V}$ fails to satisfy the $(\mathbf{1}, \mathbf{2})$ transfer principle. By Lemma 5.3, some finite algebra $\mathbf{A}$ in $\mathcal{V}$ has a three-element interval $\gamma \prec \alpha \prec \beta$ of type $(\mathbf{1}, \mathbf{2})$ in $\mathbf{Con\,A}$; and without losing generality, we can assume that $\gamma = 0_A$. Let N be an $\langle \alpha, \beta \rangle$-trace and pick $0, 1 \in N$ such that $\langle 0, 1 \rangle \in \beta - \alpha$. Let $\theta = \text{Cg}^{\mathbf{A}}(0, 1)$. Then $\theta = \beta$ since $0_A \neq \theta \neq \alpha$ and $\theta \leq \beta$. Thus $\alpha \leq \text{Cg}^{\mathbf{A}}(0, 1)$, and by Lemma 8.5 it follows that $\mathcal{V}$ is unstructured.

If $\mathcal{V}$ fails to satisfy the $(\mathbf{2}, \mathbf{1})$ transfer principle, a similar proof using Lemma 8.8 shows that $\mathcal{V}$ is unstructured. □

Chapter 9

Consequences of the transfer principles

Throughout this chapter we assume that $\mathcal{V}$ is a locally finite Abelian variety. We will show that if $\mathcal{V}$ satisfies the $(\mathbf{1},\mathbf{2})$ transfer principle, then the subvariety $\mathcal{V}_1 = V(\mathcal{S}_1)$ defined in Chapter 1 is strongly Abelian. On the other hand, if $\mathcal{V}$ is assumed to satisfy the $(\mathbf{2},\mathbf{1})$ transfer principle, then it will be shown that the subvariety $\mathcal{V}_2 = V(\mathcal{S}_2)$ is affine. Thus if $\mathcal{V}$ is structured, then we will be able to conclude with the help of Theorem 8.9 and Theorem 1.3 that $\mathcal{V}$ is the join of an affine subvariety with a strongly Abelian one.

THEOREM 9.1 *Let $\mathcal{V}$ be a locally finite Abelian variety that satisfies the $(\mathbf{1},\mathbf{2})$ transfer principle. Then the subvariety $\mathcal{V}_1 = V(\mathcal{S}_1)$ is strongly Abelian.*

PROOF. Using the $(\mathbf{1},\mathbf{2})$ transfer principle, it follows that every finite subdirectly irreducible algebra in $\mathcal{V}$ with type $\mathbf{1}$ monolith is strongly solvable (i.e., has type set equal to $\{\mathbf{1}\}$). Thus $\operatorname{typ}\{\mathcal{S}_1\} = \{\mathbf{1}\}$, and so by Theorem 0.36, $\mathcal{S}_1$ is a class of strongly solvable algebras. Then by Theorem 0.15 we conclude that $\mathcal{V}_1$ is a locally strongly solvable variety. Finally, using Corollary 7.2, we get that $\mathcal{V}_1$ is indeed strongly Abelian. □

In order to show that $\mathcal{V}_2$ is affine if the $(\mathbf{2},\mathbf{1})$ transfer principle holds in $\mathcal{V}$, we will show that $\operatorname{typ}\{\mathcal{V}_2\} = \{\mathbf{2}\}$. For the next three lemmas we assume that the $(\mathbf{2},\mathbf{1})$ transfer principle holds in $\mathcal{V}$.

LEMMA 9.2 *Let $\mathbf{A}$ be a finite solvable algebra and let a, $b \in A$ be distinct. If for all $p \in \operatorname{Pol}_2 \mathbf{A}$ the implication*

$$p(a,a) = p(b,b) \rightarrow p(a,a) = p(b,a)$$

holds, then $\mathbf{1} \in \text{typ}\{\mathbf{A}\}$.

PROOF. Suppose not and let $\beta = \text{Cg}^{\mathbf{A}}(a,b)$. Choose $\delta \in \text{Con}\,\mathbf{A}$ with $\delta \prec \beta$. Then $\text{typ}(\delta,\beta) = \mathbf{2}$. Choose $V \in M_{\mathbf{A}}(\delta,\beta)$ and $d(x,y,z) \in \text{Pol}_3\mathbf{A}$ such that the restriction of d to the body of V is Maltsev. Such a polynomial operation exists according to Lemma 0.38. Since $\langle a,b\rangle \in \beta - \delta$, we can find $f \in \text{Pol}_1\mathbf{A}$ with $f(A) \subseteq V$ and $\langle f(a), f(b)\rangle \in \beta - \delta$.

Let $p(x,y) = d(f(b), f(x), f(y)) \in \text{Pol}_2\mathbf{A}$. Then

$$p(a,a) = d(f(b), f(a), f(a)) = f(b) = d(f(b), f(b), f(b)) = p(b,b).$$

Thus by our hypotheses, $p(a,a) = p(b,a)$. But

$$p(b,a) = d(f(b), f(b), f(a)) = f(a),$$

contradicting $f(a) \neq f(b)$. The contradiction proves the lemma. □

LEMMA 9.3 *Let* $\mathbf{A}$ *be a finite algebra in* $\mathcal{V}$ *with* $\text{typ}\{\mathbf{A}\} = \{\mathbf{2}\}$. *Then* $\text{typ}\{\mathcal{S}(\mathbf{A})\} = \{\mathbf{2}\}$.

PROOF. Suppose for some subalgebra $\mathbf{B}$ of $\mathbf{A}$ we can find $\alpha \prec \beta$ in $\text{Con}\,\mathbf{B}$ with $\text{typ}(\alpha,\beta) = \mathbf{1}$. Then using the $(\mathbf{2},\mathbf{1})$ transfer principle we may assume that $\alpha = 0_B$.

Choose a, $b \in B$ with $\langle a,b\rangle \in \beta - 0_B$. Suppose that $p \in \text{Pol}_2\mathbf{A}$ and $p(a,a) = p(b,b)$. We will show that this implies that $p(a,a) = p(b,a)$, and so conclude by Lemma 9.2 that $\mathbf{1} \in \text{typ}\{\mathbf{A}\}$, contradicting our hypotheses.

Since $p \in \text{Pol}_2\mathbf{A}$, then for some $k \in \omega$, some $k+2$-ary term operation $t(x,y,\bar{z})$ and some $\bar{c} \in A$ we have $p(x,y) = t(x,y,\bar{c})$ for all x, $y \in A$. Choose any k-tuple $\bar{d}$ from B.

Since $p(a,a) = p(b,b)$ then $t(a,a,\bar{c}) = t(b,b,\bar{c})$, and so $t(a,a,\bar{d}) = t(b,b,\bar{d})$ since $\mathbf{A}$ is Abelian. Now since $\mathbf{B}$ is a subalgebra of $\mathbf{A}$ and a, b, $\bar{d} \in B$, we conclude that $t(a,a,\bar{d}) \in B$ too. But then since $\text{typ}(0_B,\beta) = \mathbf{1}$ and $\langle a,b\rangle \in \beta$ we conclude, using Theorem 0.36, that $t(a,a,\bar{d}) = t(b,a,\bar{d})$ in $\mathbf{B}$. Consequently, it follows that $t(a,a,\bar{c}) = t(b,a,\bar{c})$, i.e., that $p(a,a) = p(b,a)$, since $\mathbf{A}$ is Abelian. □

LEMMA 9.4 *Let* $\mathbf{A}_0$, $\mathbf{A}_1 \in \mathcal{V}$, *both finite. If* $\text{typ}\{\mathbf{A}_i\} = \{\mathbf{2}\}$ *for* $i = 0, 1$ *then* $\text{typ}\{\mathbf{A}_0 \times \mathbf{A}_1\} = \{\mathbf{2}\}$.

PROOF. Suppose not and let $\mathbf{C} = \mathbf{A}_0 \times \mathbf{A}_1$. Choose $\alpha \in \text{Con}\,\mathbf{C}$ a cover of 0_C with $\text{typ}(0_C,\alpha) = \mathbf{1}$. Let η_i be the kernel of the projection map from $\mathbf{C}$ onto $\mathbf{A}_i$ for $i = 0, 1$. Then either $\alpha \wedge \eta_0 = 0_C$ or $\alpha \wedge \eta_1 = 0_C$ since $\eta_0 \wedge \eta_1 = 0_C$. Without loss of generality, we may assume that $\alpha \wedge \eta_1 = 0_C$.

Thus $\eta_1 < \eta_1 \vee \alpha$. Choose $\delta \in \text{Con}\,\mathbf{C}$ with $\eta_1 \leq \delta \prec \eta_1 \vee \alpha$. Then $\alpha \vee \delta = \eta_1 \vee \alpha$, and $\alpha \wedge \delta = 0_C$, and so by Lemma 0.37, $\text{typ}(\delta, \eta_1 \vee \alpha) =$

$\operatorname{typ}(0_C, \alpha) = \mathbf{1}$. Since $\eta_1 \leq \delta \prec \eta_1 \vee \alpha$, we conclude that $\mathbf{1} \in \operatorname{typ}\{\mathbf{C}/\eta_1\} = \operatorname{typ}\{\mathbf{A}_1\}$, a contradiction. □

THEOREM 9.5 *Let $\mathcal{V}$ be a locally finite Abelian variety that satisfies the $(\mathbf{2},\mathbf{1})$ transfer principle. Then the subvariety $\mathcal{V}_2 = \boldsymbol{V}(\mathcal{S}_2)$ is affine.*

PROOF. As in Theorem 9.1, it easily follows from the $(\mathbf{2},\mathbf{1})$ transfer principle that $\operatorname{typ}\{\mathcal{S}_2\} = \{\mathbf{2}\}$. Now by Lemmas 9.3 and 9.4 it follows that $\operatorname{typ}\{\boldsymbol{S}\boldsymbol{P}_{\mathrm{fin}}(\mathcal{S}_2)\} = \{\mathbf{2}\}$ and so $\operatorname{typ}\{\mathcal{V}_2\} = \operatorname{typ}\{\boldsymbol{H}\boldsymbol{S}\boldsymbol{P}_{\mathrm{fin}}(\mathcal{S}_2)\} = \{\mathbf{2}\}$. Therefore by Theorem 0.41, $\mathcal{V}_2$ is an affine variety. □

THEOREM 9.6 *Let $\mathcal{V}$ be a locally finite, structured, Abelian variety. Then $\mathcal{V}_1$ is a strongly Abelian variety and $\mathcal{V}_2$ is an affine variety, and $\mathcal{V} = \mathcal{V}_1 \vee \mathcal{V}_2$.*

PROOF. This follows immediately from Theorems 8.9, 9.1 and 9.5. □

Chapter 10

Three interpretations

Three different, but related, constructions are required to prove Lemmas 8.4 and 8.6. Given a finite algebra $\mathbf{A}$ that satisfies the hypotheses of either of these lemmas, we will construct an interpretation of the class of all graphs into $\boldsymbol{P}_s(\mathbf{A})$. These constructions were inspired indirectly by A. P. Zamyatin [1978a], through the work of Burris and McKenzie [1981]. We will first restate Lemma 8.4 and then proceed to prove it.

Lemma 8.4 *Suppose that* $\mathrm{typ}(0_A, \alpha) = \mathbf{1}$, $\mathrm{typ}(\alpha, \beta) = \mathbf{2}$ *and* $\boldsymbol{V}(\mathbf{A})$ *is Abelian. If there is some* $U \in M_{\mathbf{A}}(0_A, \alpha)$ *and a* $\langle 0_A, \alpha\rangle$*-trace* $M \subseteq U$ *with* $M \neq M^{\beta} \cap U$ *then* $\boldsymbol{P}_s(\mathbf{A})$ *is unstructured.*

Let $\mathbf{A}$, U, M, α, β satisfy the hypotheses of this lemma. Since $M \neq M^{\beta} \cap U$ then using a standard argument from tame congruence theory we can find $V \in M_{\mathbf{A}}(\alpha, \beta)$ and $N \subseteq V$, an $\langle \alpha, \beta\rangle$-trace, such that $V \subseteq U$ and $M \cap N \neq \emptyset$. By Lemma 8.1, it follows that $M \cap N$ has precisely one element. Choose $0 \in M \cap N$, $1 \in N - M$ and $0' \in M - N$.

LEMMA 10.1 *Let* $t(x_1, \ldots, x_k)$ *be a* k*-ary polynomial operation of* $\mathbf{A}$ *such that* $t(A^k) \subseteq U$ *and for all* $a_1, \ldots, a_k \in A$ *and all* $i \leq k$ *the unary polynomial operations* $t(a_1, a_2, \ldots, a_{i-1}, x, a_{i+1}, \ldots, a_k)$ *are nonconstant on* $M \cup N$. *Then either* $t(a_1, a_2, \ldots, a_{i-1}, x, a_{i+1}, \ldots, a_k)$ *is constant on* M *for all* $a_1, \ldots, a_k \in A$ *and all* $i \leq k$, *or* $k = 1$ *and* $t(x_1)$ *maps* U *bijectively onto* U.

PROOF. If this lemma is false, then we can find $t(x, y) \in \mathrm{Pol}_2\mathbf{A}$ such that $t(A \times A) \subseteq U$, $t(x, c)$ and $t(c, x)$ are both nonconstant on $M \cup N$ for all $c \in A$ and for some $a \in A$, $t(x, a)$ is nonconstant on M. We will show that this leads to a contradiction.

Since $t(x,a)$ is nonconstant on M, and M is a $\langle 0_A, \alpha\rangle$-trace contained in U, then $t(U,a) = U$. But then $t(U,c) = U$ for all $c \in A$ (we are using the fact that $t(A \times A) \subseteq U$, $U \in M_{\mathbf{A}}(0_A, \alpha)$ and $\mathbf{A}$ is Abelian). We note that $t(M,c)$ is a $\langle 0_A, \alpha\rangle$-trace contained in U for all $c \in A$. Since $\operatorname{typ}(0_A, \alpha) = \mathbf{1}$, then by Lemma 0.40, $t|_{M^2}$ is essentially unary, and so for all $c \in A$, $t(c,x)$ is constant on M. It follows that $t(c, \alpha|_U) \subseteq 0_A$ for all c, since $U \in M_{\mathbf{A}}(0_A, \alpha)$.

If $t(c, \beta|_V) \subseteq \alpha$ for some $c \in A$, then $t(c,N) \subseteq M'$ for some $\langle 0_A, \alpha\rangle$-trace $M' \subseteq U$ since $t(c,x)$ is nonconstant on N. But then by Lemma 8.2 we conclude that $\boldsymbol{HS}(\mathbf{A}^2)$ is not Abelian, which is contrary to our assumption that $\boldsymbol{V}(\mathbf{A})$ is Abelian. Thus for all $c \in A$, $t(c, \beta|_N) \not\subseteq \alpha$ and so $t(c,N)$ is an $\langle \alpha, \beta\rangle$-trace contained in U.

Since $t(U,a) = U$ for all $a \in A$, then $t(x,a)$ permutes the $\beta|_U$ equivalence classes. Thus

$$t((0/\beta) \cap U, 1) = (t(0,1)/\beta) \cap U = (t(0,0)/\beta) \cap U = t((0/\beta) \cap U, 0)$$

since $\langle 0,1\rangle \in \beta$. Thus we can find $c \in U$ such that $\langle 0,c\rangle \in \beta$ and $t(c,1) = t(0,0)$. Now, $\langle 0,c\rangle \notin \alpha$, for if not then

$$\langle t(c,1), t(0,1)\rangle \in \alpha, \quad \text{implying} \quad \langle t(0,0), t(0,1)\rangle \in \alpha.$$

This contradicts the fact that $t(0, \beta|_N) \not\subseteq \alpha$, established earlier.

Thus $\langle 0,c\rangle \in \beta - \alpha$ and so we can find $h(x) \in \operatorname{Pol}_1 \mathbf{A}$ such that $h(A) \subseteq V$ and $\langle h(0), h(c)\rangle \in \beta - \alpha$. We may choose h so that $h(0)$, $h(c) \in N$, since all $\langle \alpha, \beta\rangle$-traces in V are polynomially isomorphic in $\mathbf{A}$. Since $\operatorname{typ}(\alpha, \beta) = \mathbf{2}$ and $\operatorname{typ}(0_A, \alpha) = \mathbf{1}$ then using Lemma 8.1 and Lemma 0.32, we have that the induced algebra $\mathbf{A}|_N$ is polynomially equivalent to a one dimensional vector space over some finite field. So we may choose $h(x)$ so that $h(0) = 0$ and $h(c) = 1$.

Let $g(x) = t(x, h(x)) \in \operatorname{Pol}_1 \mathbf{A}$. Then

$$g(0) = t(0, h(0)) = t(0,0)$$

and

$$g(0') = t(0', h(0')) = t(0', h(0)) = t(0', 0) \neq g(0),$$

since $t(x,0)$ is a bijection on U, $\langle 0, 0'\rangle \in \alpha$ and $t(0', \alpha|_U) \subseteq 0_A$. Thus $g(\alpha|_U) \not\subseteq 0_A$ and so by $\langle 0_A, \alpha\rangle$-minimality we conclude that $g(x)$ is one to one on U. But

$$g(c) = t(c, h(c)) = t(c,1) = t(0,0) = g(0)$$

which is a contradiction. This completes the proof of the lemma. □

Proof of Lemma 8.4. Let $\mathbf{G} = \langle G, E\rangle$ be any graph in which $|G| > 2$. Let p_1, p_2 be two distinct points not in G and let $X = G \cup \{p_1, p_2\}$. We proceed to construct a subalgebra of $\mathbf{A}^X$ from which we can recover $\mathbf{G}$ using a suitable interpretation scheme. Since the class of all graphs interprets into the class of all graphs with at least three vertices (and since our interpretation scheme will not depend on $\mathbf{G}$) then this will suffice to prove the lemma.

For $v \in G$, let $f_v : X \to \{0, 0', 1\}$ be defined as

$$f_v(x) = \begin{cases} 0' & \text{if } x = p_2 \\ 1 & \text{if } x = v \\ 0 & \text{otherwise,} \end{cases}$$

and for $e \in E$ and $i = 1, 2$ let $f_e^i : X \to \{0, 0', 1\}$ be defined as

$$f_e^i(x) = \begin{cases} 0' & \text{if } x = p_i \\ 1 & \text{if } x \in e \\ 0 & \text{otherwise.} \end{cases}$$

Let

$$G^* = \{f_v : v \in G\},$$
$$E^* = \{f_e^i : e \in E,\ i = 1, 2\}$$

and let

$$N^* = \{f \in N^X : f(p_1) = f(p_2)\}.$$

Recall that for $a \in A$, we use $\hat{a}$ to denote the constant function from X to A with value a. We define

$\mathbf{D}$ is the subalgebra of $\mathbf{A}^X$
generated by the set $G^* \cup E^* \cup N^* \cup \{\hat{a} : a \in A\}$.

Notice that $\mathbf{D}$ is a diagonal subalgebra of $\mathbf{A}^X$. Also, for $\mu \in \mathbf{D}$ we have $\langle \mu(p_1), \mu(p_2)\rangle \in \alpha$, since the generators satisfy this.

We will now set out to define the sets G^* and E^* in $\mathbf{D}$. Choose $e_1, e_2 \in \mathrm{E}(\mathbf{A})$ satisfying $e_1(A) = U$ and $e_2(A) = V$. These functions are defined in $\mathbf{A}$ by first order formulas with parameters, and the same formulas (with different parameters) define the functions $e_1^{(X)}|_D$ and $e_2^{(X)}|_D$, which we denote again simply by e_1 and e_2. Of course, $D(U) = U^X \cap D$ is the range of the polynomial e_1 of $\mathbf{D}$, and a similar statement holds for $D(V)$. Thus

$D(U)$ and $D(V)$ are definable in $\mathbf{D}$
and $G^* \cup E^* \cup N^* \subseteq D(U)$.

Now let $\mu \in D(U)$ be an arbitrary element. If μ is not a constant then for some k and $t(x_1, \ldots, x_k) \in \mathrm{Pol}_k\mathbf{A}$ we can find

$$\nu_1, \ldots, \nu_k \in G^* \cup E^* \cup N^*$$

such that $t(A^k) \subseteq U$, $\mu = t^{(X)}(\nu_1, \ldots, \nu_k)$ and for all $a_1, \ldots, a_k \in A$ and $i \leq k$, $t(a_1, \ldots, a_{i-1}, x, a_{i+1}, \ldots, a_k)$ is nonconstant on $M \cup N$. Then by Lemma 10.1, we conclude that either $t(a_1, \ldots, a_{i-1}, x, a_{i+1}, \ldots, a_k)$ is constant on M for all $a_1, \ldots, a_k \in A$ and $i \leq k$, or $k = 1$ and $t(x_1)$ maps U bijectively onto U. If the former holds then we call t a **collapsing function** since $\alpha|_U$ is collapsed into 0_U at each variable.

For $k \in \omega$ and $t(x_1, \ldots, x_k) \in \mathrm{Pol}_k\mathbf{A}$, if t is a collapsing function, let $N(t) = t(N^k)$. Since A is finite, then there exists finitely many collapsing functions $t_1, \ldots, t_m$ such that for all collapsing functions t, $N(t) = N(t_i)$ for some $i \leq m$. Let t_i have arity n_i.

Claim 1. There is a first order formula (with parameters) $\mathrm{Coll}(x)$ such that for $\mu \in D$, $\mathbf{D} \models \mathrm{Coll}(\mu)$ if and only if $\mu \in D(U)$ and either μ is constant or $\mu = t^{(X)}(\nu_1, \ldots, \nu_k)$ for some $k \in \omega$, some nonconstant generators $\nu_1, \ldots, \nu_k$ and some collapsing function $t \in \mathrm{Pol}_k\mathbf{A}$.

To prove the claim let

$$\mathrm{Con}(x) \overset{\text{def}}{\longleftrightarrow} \bigvee_{a \in A} x \approx \hat{a}.$$

Now, if $\mu = t^{(X)}(\nu_1, \ldots, \nu_k)$ for some nonconstant generators $\nu_1, \ldots, \nu_k$ and some collapsing function t, then we may choose the $\nu_i \in N^*$, since for all $a_1, \ldots, a_k \in A$ and $i \leq k$,

$$t(a_1, \ldots, a_{i-1}, 0, a_{i+1}, \ldots, a_k) = t(a_1, \ldots, a_{i-1}, 0', a_{i+1}, \ldots, a_k).$$

Choose $i \leq m$ such that $N(t) = N(t_i)$. We will construct $\mu_1, \ldots, \mu_{n_i} \in N^*$ such that $\mu = t_i{}^{(X)}(\bar{\mu})$.

Since $N(t) = N(t_i)$ then for $x \in X - \{p_2\}$ we can find $u_1(x), \ldots, u_{n_i}(x) \in N$ such that

$$\mu(x) = t(\nu_1(x), \ldots, \nu_k(x)) = t_i(u_1(x), \ldots, u_{n_i}(x)).$$

For $j \leq n_i$ and $x \in X - \{p_2\}$, let $\mu_j(x) = u_j(x)$ and let $\mu_j(p_2) = u_j(p_1)$. Then $\mu_j \in N^*$ and $\mu = t_i{}^{(X)}(\mu_1, \ldots, \mu_{n_i})$.

We take for $\mathrm{Coll}(x)$ a first order formula equivalent to

$$\begin{array}{c} x \in D(U) \text{ and either } \mathrm{Con}(x) \text{ or} \\ \text{for some } j \in \{1, \ldots, m\} \text{ and elements} \\ y_1, \ldots, y_{n_j} \text{ in } D(V), x = t_j^{(X)}(y_1, \ldots, y_{n_j}). \end{array} \tag{10.1}$$

Note that the polynomial operation $t_j^{(X)}$ of $\mathbf{D}$ is definable in the same way that $e_1^{(X)}$ is definable; therefore (10.1) is equivalent to a first order formula in which the $\hat{a}$ ($a \in A$) appear as parameters. We have just shown that if $\mu = t^{(X)}(\nu_1, \ldots, \nu_k)$ for some nonconstant generators $\nu_1, \ldots, \nu_k$, and some collapsing function t, then $\mathbf{D} \models \mathrm{Coll}(\mu)$.

Before beginning the proof of the converse, we make an observation which will be needed here and later in the proof. If we had $U = V$ then also $M \subseteq N$ since $\alpha < \beta$. However, $\alpha|_N = 0_N$ by Lemma 8.1; hence V is a proper subset of U. It follows that if $g \in \mathrm{Pol}_1 \mathbf{A}$ and $g(U) \subseteq V$ then g is a collapsing function, i.e., $g(\alpha|_U) \subseteq 0_A$.

Now suppose that $\mathbf{D} \models \mathrm{Coll}(\mu)$ and μ is nonconstant. As we have already shown, either $\mu = g^{(X)}(\nu)$ for some nonconstant generator ν and some $g \in \mathrm{Pol}_1 \mathbf{A}$ with $g(U) = U$, or $\mu = t_i{}^{(X)}(\nu_1, \ldots, \nu_{n_i})$ for some nonconstant generators $\nu_1, \ldots, \nu_{n_i}$ and some $i \leq m$.

If the latter holds, then we are fine, but if the former holds then we have two cases to consider. If the generator ν is in N^*, then setting $t(x) = g(e_2(x))$, we see that $\mu = t^{(X)}(\nu)$, and that $t(x)$ is a collapsing function, so there is nothing more to prove.

If the generator ν is in $G^* \cup E^*$, then $\langle \mu(p_1), \mu(p_2) \rangle \in \alpha - 0_A$ since ν satisfies this, and $g(U) = U$. Since $\mathbf{A}$ is Abelian and $\mathrm{typ}(\alpha, \beta) = \mathbf{2}$, then by Lemma 0.38 the tail of V is empty and so by Lemma 8.1 we have $\alpha|_V = 0_V$. Thus for all $\eta \in D(V)$, $\eta(p_1) = \eta(p_2)$. But then $\mathrm{Coll}(\mu)$ is impossible, for if $\mathbf{D} \models \mathrm{Coll}(\zeta)$, then $\zeta(p_1) = \zeta(p_2)$. Thus we have proved Claim 1. Observe that the above proof shows that $\mathrm{Coll}^{\mathbf{D}}$ is just the set of all $\mu \in D(U)$ satisfying $\mu(p_1) = \mu(p_2)$.

We define

$$\mathrm{Gen}(x) \overset{\mathrm{def}}{\longleftrightarrow} x \in D(U) \wedge \neg\, \mathrm{Coll}(x).$$

By Claim 1 it follows that

$$\begin{aligned} &\mathbf{D} \models \mathrm{Gen}(\mu) \text{ iff} \\ &\mu \in D(U) \text{ and } \mu = g^{(X)}(f_v) \text{ or } \mu = g^{(X)}(f_e^i) \end{aligned}$$

for some $v \in G$, $e \in E$, $i = 1, 2$ and $g \in \mathrm{Pol}_1 \mathbf{A}$ with $g(A) \subseteq U$ and $g(U) = U$. Let

$$S = \{g \in \mathrm{Pol}_1 \mathbf{A} : g(A) \subseteq U \text{ and } g(U) = U\}$$

and let $\mathrm{Eq}(x, y)$ be a first order formula equivalent to

$$\begin{aligned} &\{x, y\} \subseteq D(U) \text{ and } x = g^{(X)}(y) \\ &\text{for some } g \in S. \end{aligned}$$

Evidently, $\mathrm{Eq}^{\mathbf{D}}$ is an equivalence relation on $D(U)$. We define the symmetric relation $x \sim y$ as follows:

$$x \sim y \overset{\text{def}}{\longleftrightarrow}$$

$$(\exists u, v)\,[\,\mathrm{Eq}(u,x) \wedge \mathrm{Eq}(v,y) \wedge \neg\,\mathrm{Con}(e_2(u)) \wedge e_2(u) \approx e_2(v)]\,.$$

Claim 2. Let $\mu, \nu \in \mathbf{D}$ be such that

$$\mathbf{D} \models \mathrm{Gen}(\mu) \;\wedge\; \mathrm{Gen}(\nu).$$

If $\mu = h^{(X)}(f_v)$ for some $v \in G$ and $h \in S$, then

$$\mathbf{D} \models \mu \sim \nu \leftrightarrow \mathrm{Eq}(\mu, \nu).$$

If on the other hand, $\mu = h^{(X)}(f_e^i)$ for some $e \in E$, $i \leq 2$, and $h \in S$, then

$$\mu \sim \nu \text{ iff } \nu = k^{(X)}(f_e^j) \text{ for some } j \leq 2 \text{ and } k \in S.$$

To begin the proof of this claim, suppose that $\mu = h^{(X)}(f_v)$ for $h \in S$ and $v \in G$ and suppose $\mu \sim \nu$. Since $\mathrm{Gen}(\nu)$ then $\nu = k^{(X)}(f_w)$ or $k^{(X)}(f_e^i)$ for some $k \in S$, some $w \in G$ or $e \in E$, and $i \leq 2$. Let's consider the case when $\nu = k^{(X)}(f_e^i)$. Since $\mu \sim \nu$ then for some γ, $\delta \in D$,

$$\mathrm{Eq}(\gamma, \mu) \wedge \mathrm{Eq}(\delta, \nu) \wedge \neg\,\mathrm{Con}({e_2}^{(X)}(\gamma)) \wedge {e_2}^{(X)}(\gamma) = {e_2}^{(X)}(\delta)$$

and so for some l_1, $l_2 \in S$, we have

$${e_2}^{(X)}({l_1}^{(X)}(f_v)) = {e_2}^{(X)}({l_2}^{(X)}(f_e^i))$$

and ${e_2}^{(X)}({l_1}^{(X)}(f_v))$ is nonconstant. But this equation implies

$$e_2(l_1(f_v(p_1))) = e_2(l_2(f_e^i(p_1)))$$

and so

$$e_2(l_1(0')) = e_2(l_1(0)) = e_2(l_2(0)) = e_2(l_2(0')).$$

We have used the fact that $e_2(\alpha|_U) \subseteq 0_U$. Now choose $w \in G$ such that $w \in e - \{v\}$. Then $e_2(l_1(f_v(w))) = e_2(l_2(f_e^i(w)))$ implies $e_2(l_1(0)) = e_2(l_2(1))$. But then ${e_2}^{(X)}({l_2}^{(X)}(f_e^i))$ is constant, which is contrary to $\neg\mathrm{Con}({e_2}^{(X)}(\gamma))$. Thus $\nu = k^{(X)}(f_e^i)$ is impossible.

Using a similar argument, one can show that if $\nu = k^{(X)}(f_w)$, then $v = w$, and so $\mathbf{D} \models \mathrm{Eq}(\mu, \nu)$. The converse is easy to prove.

The remaining case is when $\mu = h^{(X)}(f_e^i)$ for some e, i and h. Since $\sim$ is a symmetric relation, then the above argument shows that $\nu = k^{(X)}(f_{e'}^j)$ for some $k \in S$, $e' \in E$ and $j \leq 2$. To show that $e' = e$, one can argue as above. For the converse, it is not difficult to see that

$$\mathbf{D} \models f_e^i \sim f_e^j \quad \text{for all } e \in E \text{ and } i, j \leq 2 \,.$$

This ends the proof of Claim 2. Notice that since $|G| > 2$ we have $\neg\ \mathrm{Eq}(f_e^1, f_e^2)$; so that Claim 2 implies that an element $\mu \in \mathrm{Gen}^{\mathbf{D}}$ is of the form $h^{(X)}(f_v)$ for some $v \in G$ and $h \in S$ iff $\mathbf{D} \models \mu \sim \nu \leftrightarrow \mathrm{Eq}(\mu, \nu)$ for all $\nu \in \mathrm{Gen}^{\mathbf{D}}$.

Since $\mathrm{typ}(\alpha, \beta) = \mathbf{2}$ and N is an $\langle \alpha, \beta \rangle$-trace with $\alpha|_N = 0_N$, then we can find a binary polynomial operation $x + y$ of $\mathbf{A}$ such that the range of $+$ is contained in V and $+$ restricted to N acts like vector space addition on $\mathbf{A}|_N$ with 0 as the additive identity. In the following discussion we will also use the symbol $+$ to denote the operation $+^{(X)}|_D$ on $\mathbf{D}$.

We are now in a position to give the interpretation scheme. Let

$$\mathrm{Un}(x) \quad \overset{\text{def}}{\longleftrightarrow} \quad \Big(\mathrm{Gen}(x) \wedge \forall y\, \big([\mathrm{Gen}(y) \wedge y \sim x] \rightarrow \mathrm{Eq}(x, y)\big)\Big)$$

and

$$\begin{aligned} \mathrm{E}(x, y) \quad \overset{\text{def}}{\longleftrightarrow} \quad & \Big[\mathrm{Un}(x) \wedge \mathrm{Un}(y) \ \wedge \neg\ \mathrm{Eq}(x, y) \ \wedge \\ & \exists a, b, c \Big(\mathrm{Gen}(a) \wedge \mathrm{Gen}(b) \wedge \mathrm{Gen}(c) \wedge \mathrm{Eq}(a, x) \wedge \mathrm{Eq}(b, y) \ \wedge \\ & \neg\ \mathrm{Un}(c) \wedge \neg\ \mathrm{Con}(e_2(c)) \wedge e_2(c) \approx e_2(a) + e_2(b)\Big)\Big]. \end{aligned}$$

It follows from Claims 1 and 2 (as noted above) that for $\mu \in D$, we have $\mathbf{D} \models \mathrm{Un}(\mu)$ iff $\mu = h^{(X)}(f_v)$ for some $h \in S$ and $v \in G$.

Claim 3. For μ, $\nu \in D$,

$$\mathbf{D} \models \mathrm{E}(\mu, \nu) \text{ if and only if } \mu = h^{(X)}(f_v) \text{ and } \nu = k^{(X)}(f_w)$$

for some h, $k \in S$ and v, $w \in G$ with $\{v, w\} \in E$.

To prove this claim, we note first that one direction is clear—that is, if $\mu = h^{(X)}(f_v)$ and $\nu = k^{(X)}(f_w)$ with h, $k \in S$ and $\{v, w\} = e \in E$, then $\mathbf{D} \models \mathrm{E}(\mu, \nu)$. In fact, setting $a = f_v$, $b = f_w$ and $c = f_e^1$, we see that

$$\mathrm{Eq}(a, \mu) \wedge \mathrm{Eq}(b, \nu) \wedge \neg\ \mathrm{Un}(c),$$

and that

$$e_2{}^{(X)}(c) = e_2{}^{(X)}(a) + e_2{}^{(X)}(b),$$

since $e_2(0) = e_2(0')$ and $0 + 1 = 1 + 0 = 1$.

Conversely, suppose that $\mathbf{D} \models \mathrm{E}(\mu, \nu)$ with $\mu = h^{(X)}(f_v)$ and $\nu = k^{(X)}(f_w)$ for some h, $k \in S$. Then for some $e \in E$, $i \leq 2$ and l_1, l_2, $l_3 \in S$ we have

$$\neg \operatorname{Con}\left(e_2{}^{(X)}(l_3{}^{(X)}(f_e^i))\right)$$

and

$$e_2{}^{(X)}(l_3{}^{(X)}(f_e^i)) = e_2{}^{(X)}(l_2{}^{(X)}(f_v)) + e_2{}^{(X)}(l_1{}^{(X)}(f_w)).$$

If $e \neq \{v, w\}$ then choose $r \in e - \{v, w\}$. Then evaluating the above equality at the coordinate r we find that $e_2(l_3(1)) = e_2(l_2(0)) + e_2(l_1(0))$. But evaluating the equality at the coordinate p_i we get

$$e_2(l_3(0)) = e_2(l_2(0)) + e_2(l_1(0)),$$

implying

$$e_2(l_3(0')) = e_2(l_3(0)) = e_2(l_3(1)).$$

Thus $\mathbf{D} \models \operatorname{Con}\left(e_2{}^{(X)}(l_3{}^{(X)}(f_e^i))\right)$ which is contrary to our assumptions. Hence we conclude that $e = \{v, w\} \in E$.

The above claims have established that the binary structure

$$\langle \mathrm{Un}^{\mathbf{D}}, \mathrm{E}^{\mathbf{D}} \rangle / \mathrm{Eq}^{\mathbf{D}}$$

is isomorphic to our original graph $\mathbf{G} = \langle G, E \rangle$ via the map which sends $f_v/\mathrm{Eq}^{\mathbf{D}}$ to v in G.

Thus we have successfully interpreted (via parameters) the graph $\mathbf{G}$ into an algebra in $\boldsymbol{P}_s(\mathbf{A})$. Since the interpretation scheme is independent of the choice of $\mathbf{G}$, we can conclude that $\boldsymbol{P}_s(\mathbf{A})$ is unstructured. □

To prove Lemma 8.6 we will have to consider two different cases. Some of the notational conventions established in the proof of Lemma 8.4 will be observed in the following.

Lemma 8.6 *Suppose that* $\mathrm{typ}(0_A, \alpha) = \mathbf{2}$, $\mathrm{typ}(\alpha, \beta) = \mathbf{1}$, *and* $V(\mathbf{A})$ *is Abelian. Let* N *be an* $\langle \alpha, \beta \rangle$*-trace and* M *a* $\langle 0_A, \alpha \rangle$*-trace, and suppose there exists some* $f \in \mathrm{Pol}_1 \mathbf{A}$ *such that* $f(N) \subseteq M$ *and* f *is nonconstant on* N. *Then the class* $\boldsymbol{P}_s(\mathbf{A})$ *is unstructured.*

Let $\mathbf{A}$, M, N, α, β satisfy the hypotheses of Lemma 8.6. Choose an $\langle \alpha, \beta \rangle$-minimal set V which contains N. Let $f \in \mathrm{Pol}_1 \mathbf{A}$ be such that $f(N) \subseteq M$ and f is nonconstant on N. Then $f(\beta|_V) \not\subseteq 0_A$ and $f(\beta|_V) \subseteq \alpha$ since $V \in M_{\mathbf{A}}(\alpha, \beta)$.

The first case to be considered is when we can find such an f with $f(\alpha|_V) \subseteq 0_A$.

LEMMA 10.2 *Let* $\mathbf{A}$, M, N, V, α, β *be as above and suppose that for some* $f \in \mathrm{Pol}_1 \mathbf{A}$ *we have* $f(N) \subseteq M$, f *is nonconstant on* N *and* $f(\alpha|_V) \subseteq 0_A$. *Then* $\mathbf{P}_s(\mathbf{A})$ *is unstructured.*

PROOF. Choose $U \in M_{\mathbf{A}}(0_A, \alpha)$ such that $M \subseteq U$. Since $f(N) \subseteq M$, we may assume that $f(A) \subseteq U$. We have noted above that $f(\beta|_V) \subseteq \alpha$ and $f(\beta|_V) \not\subseteq 0_A$. Recall that by assumption we also have $f(\alpha|_V) \subseteq 0_A$. Choose a, $b \in N$ such that $f(a) \neq f(b)$ and let $0 = f(a)$ and $1 = f(b)$.

Let $\mathbf{G} = \langle G, E \rangle$ be any graph with at least three vertices. Our construction is similar to the one used in the proof of Lemma 8.4. Choose two distinct points p_1, p_2 not in G and let $X = G \cup \{p_1, p_2\}$. We can now define the algebra from which we can recover $\mathbf{G}$.

For $v \in G$, let $h_v : X \to \{a, b\}$ be defined as

$$h_v(x) = \begin{cases} b & \text{if } x = v \\ a & \text{otherwise,} \end{cases}$$

and for $e \in E$, let $h_e^1, h_e^2 : X \to \{a, b\}$ be defined as

$$h_e^1(x) = \begin{cases} b & \text{if } x \in e \\ a & \text{otherwise} \end{cases}$$

and

$$h_e^2(x) = \begin{cases} b & \text{if } x \in e \text{ or } x = p_1 \\ a & \text{otherwise.} \end{cases}$$

Finally let $\chi : X \to \{a, b\}$ be defined as

$$\chi(x) = \begin{cases} b & \text{if } x = p_1 \\ a & \text{otherwise.} \end{cases}$$

Let

$$G^* = \{h_v : v \in G\},$$
$$E^* = \{h_e^i : e \in E, i = 1, 2\}$$

and define

$\mathbf{D}$ is the subalgebra of $\mathbf{A}^X$
generated by $G^* \cup E^* \cup \{\chi\} \cup \{\hat{a} : a \in A\}$.

Again, observe that $\mathbf{D}$ is a diagonal subdirect power of $\mathbf{A}$ and that for all $\mu \in D$ and all $x_1, x_2 \in X$, we have $\langle \mu(x_1), \mu(x_2) \rangle \in \beta$ since this is true of the generators.

Choose $e_1, e_2 \in \mathrm{E}(\mathbf{A})$ with $e_1(A) = U$ and $e_2(A) = V$. Let

$$S = \{p \in \mathrm{Pol}_1 \mathbf{A} : p(U) = U\}.$$

Proceeding as in the proof of Lemma 8.4, we choose formulas $\mathrm{Con}(x)$ and $\mathrm{Eq}(x,y)$ so that $\mathbf{D} \models \mathrm{Con}(\mu)$ if and only if $\mu = \hat{a}$ for some $a \in A$, and $\mathbf{D} \models \mathrm{Eq}(\mu,\nu)$ is equivalent to

$$\{\mu,\nu\} \subseteq D(U) \text{ and } \mu = g^{(X)}(\nu) \\ \text{for some } g \in S.$$

Then as in the earlier proof, $\mathrm{Eq}^{\mathbf{D}}$ is an equivalence relation on $D(U)$.

For $\mu,\nu \in \mathbf{D}$ we say that μ and ν **have the same shape**, and write $\mathrm{sh}(\mu) = \mathrm{sh}(\nu)$, if and only if for all $x_1, x_2 \in X$, $\mu(x_1) = \mu(x_2)$ if and only if $\nu(x_1) = \nu(x_2)$. In the case that $\mu,\nu \in D(U)$ and are both two-valued, we will prove that

$$\mathbf{D} \models \mathrm{Eq}(\mu,\nu) \text{ if and only if } \mathrm{sh}(\mu) = \mathrm{sh}(\nu).$$

One direction is clear, for if $p \in S$ and $p^{(X)}(\mu) = \nu$ then since p is one to one on U, we easily see that $\mathrm{sh}(\mu) = \mathrm{sh}(\nu)$.

For the converse, we first note that since $\mathrm{typ}(0_A,\alpha) = \mathbf{2}$ and $\mathbf{A}$ is Abelian, then $\mathbf{A}|_U$ is Maltsev according to Lemma 0.38. The congruences $\beta|_U$ and $\alpha|_U$ of the Maltsev algebra $\mathbf{A}|_U$ have the property that $\beta|_U$ is strongly Abelian over $\alpha|_U$ (since $\mathrm{typ}(\alpha,\beta) = \mathbf{1}$), and thus $\beta|_U = \alpha|_U$. Thus U can contain no $\langle\alpha,\beta\rangle$-minimal set; and if $\sigma \in D(U)$ then for all $x_1, x_2 \in X$, $\langle\sigma(x_1),\sigma(x_2)\rangle \in \alpha$, since $\langle\sigma(x_1),\sigma(x_2)\rangle \in \beta$.

Now suppose that μ, $\nu \in D(U)$ are two-valued and $\mathrm{sh}(\mu) = \mathrm{sh}(\nu)$. Then for some pair of $\langle 0_A,\alpha\rangle$-traces $N_1, N_2 \subseteq U$ and elements $c,d \in N_1$ and $k,l \in N_2$ we have $\mu \in \{c,d\}^X$ and $\nu \in \{k,l\}^X$. We can assume that for all $x \in X$, $\mu(x) = c$ if and only if $\nu(x) = k$. Choose $h_1 \in S$ such that $h_1(N_1) = N_2$ (such an h_1 exists by Theorem 0.33). Now since $\mathbf{A}|_{N_2}$ is polynomially equivalent to a one dimensional vector space over a finite field, then for some $h_2 \in S$ we have $h_2(h_1(c)) = k$ and $h_2(h_1(d)) = l$. This is because the action of the set of unary polynomials of a one dimensional vector space is doubly transitive on the vector space. Setting $h(x) = h_2(h_1(x))$, we see that $h \in S$ and $h^{(X)}(\mu) = \nu$. Therefore $\mathbf{D} \models \mathrm{Eq}(\mu,\nu)$.

We now can define a formula $\mathrm{Gen}(x)$ which comes close to defining the set $f^{(X)}(G^* \cup E^*)$ in $\mathbf{D}$. (Recall that f is our earlier chosen polynomial function of $\mathbf{A}$ satisfying $f(N) \subseteq M$.) Let $\mathrm{Gen}(x)$ be a formula equivalent to

$$x \in D(U) \text{ and } \neg\, \mathrm{Con}(x) \text{ and } \neg\, \mathrm{Eq}(x, f(\chi)) \\ \text{and for some } y \in D(V), \mathrm{Eq}(x, f(y)).$$

Claim 1. For $\mu \in \mathbf{D}$, $\mathbf{D} \models \mathrm{Gen}(\mu)$ if and only if $\mu \in D(U)$ and for some $\nu \in G^* \cup E^*$ we have $\mathbf{D} \models \mathrm{Eq}(\mu, f^{(X)}(\nu))$ (i.e., $\mathrm{sh}(\mu) = \mathrm{sh}(\nu)$) .

To prove this claim, suppose first that $\mathbf{D} \models \mathrm{Eq}(\mu, f^{(X)}(\nu))$ and $\nu \in G^* \cup E^*$. We see that μ is not constant, since $f^{(X)}(\nu)$ isn't and

$$\mathrm{sh}(\mu) = \mathrm{sh}(f^{(X)}(\nu)) = \mathrm{sh}(\nu) \neq \mathrm{sh}(\chi).$$

From this it follows that $\mathbf{D} \models \neg\, \mathrm{Eq}(\mu, f^{(X)}(\chi))$ and so $\mathbf{D} \models \mathrm{Gen}(\mu)$.

Conversely, suppose that $\mathbf{D} \models \mathrm{Gen}(\mu)$. Then $\mu \in D(U)$, μ is not constant, and for some $\sigma \in D(V)$ we have $\mathrm{Eq}(\mu, f^{(X)}(\sigma))$. Since $\sigma \in D(V)$ then for some $k \in \omega$, some $p(x_1, \ldots, x_k) \in \mathrm{Pol}_k \mathbf{A}$ and some nonconstant generators $\nu_1, \ldots, \nu_k$ we have $p(A^k) \subseteq V$ and $\sigma = p^{(X)}(\nu_1, \ldots, \nu_k)$.

Since μ is nonconstant and $\mathbf{D} \models \mathrm{Eq}(\mu, f^{(X)}(\sigma))$ then σ is nonconstant modulo α since $f(\alpha|_V) \subseteq 0_A$. Thus for some $i \leq k$, say $i = 1$, we have

$$p(\beta|_N, a, \ldots, a) \not\subseteq \alpha,$$

or else, because $\mathbf{A}/\alpha$ is Abelian, we would have that σ is constant modulo α.

Now by Lemma 0.40 we must have that $p|_{N^k}$ is essentially unary modulo α, since $\mathrm{typ}(\alpha, \beta) = \mathbf{1}$, and so for $i > 1$,

$$p(a, \ldots, a, \beta|_N, a, \ldots, a) \subseteq \alpha.$$

Thus for all $x \in X$ we have $\langle \sigma(x), p(\nu_1(x), a, \ldots, a)\rangle \in \alpha$. But then

$$f^{(X)}(\sigma) = f^{(X)} h^{(X)}(\nu_1)$$

where $h(x) = p(x, a, \ldots, a)$ since $f(\alpha|_V) \subseteq 0_A$. Then since ν_1 is two-valued, it follows that $f^{(X)}(\sigma)$ has at most two values. Then because $\mathrm{sh}(\mu) = \mathrm{sh}(f^{(X)}(\sigma))$ and μ is nonconstant, it follows that $f^{(X)}(\sigma)$ is two-valued and so we have that $\mathrm{sh}(f^{(X)}(\sigma)) = \mathrm{sh}(\nu_1) = \mathrm{sh}(f^{(X)}(\nu_1))$. Now it follows by an earlier remark that we have $\mathbf{D} \models \mathrm{Eq}(\mu, f^{(X)}(\nu_1))$. This concludes our proof of Claim 1.

A formula is needed to distinguish between the elements in G^* and those in E^*. Choose $d \in \mathrm{Pol}_3 \mathbf{A}$ such that $d|_U$ is a Maltsev operation on U. (See Lemma 0.38 for the existence of such a polynomial.) Let $x + y = d(x, 0, y)$. Then since $\mathbf{A}|_U$ is Abelian, the algebra $\langle U, +|_U\rangle$ is an Abelian group. As in the proof of Lemma 8.4 we will use the symbol $+$ to also denote the operation $+^{(X)}|_D$ on $\mathbf{D}$. Let $\mathrm{Edg}(x)$ be the formula

$$\mathrm{Gen}(x) \wedge \exists x', y \left(\mathrm{Gen}(x') \wedge \mathrm{Gen}(y) \wedge \mathrm{Eq}(x, x') \wedge \right.$$

$$\left. \left[(y + f^{(X)}(\chi) \approx x') \vee (x' + f^{(X)}(\chi) \approx y) \right] \right).$$

Claim 2. For $\mu \in \mathbf{D}$,

$$\mathbf{D} \models \mathrm{Edg}(\mu) \text{ if and only if } \mathbf{D} \models \mathrm{Gen}(\mu) \wedge \mathrm{Eq}(\mu, f^{(X)}(h_e^i))$$

for some $e \in E$ and $i \leq 2$.

For the proof of this claim, suppose first that

$$\mathbf{D} \models \mathrm{Gen}(\mu) \wedge \mathrm{Eq}(\mu, f^{(X)}(h_e^i)).$$

Let $\mu' = f^{(X)}(h_e^i)$ and $\nu = f^{(X)}(h_e^j)$ where $\{i, j\} = \{1, 2\}$. It follows from our earlier claims that we have

$$\mathrm{Gen}(\mu') \wedge \mathrm{Gen}(\nu) \wedge \mathrm{Eq}(\mu, \mu')$$

in $\mathbf{D}$. If $i = 1$ then

$$\mu' + f^{(X)}(\chi) = f^{(X)}(h_e^1) + f^{(X)}(\chi) = f^{(X)}(h_e^2) = \nu$$

and if $i = 2$ then

$$\nu + f^{(X)}(\chi) = f^{(X)}(h_e^1) + f^{(X)}(\chi) = f^{(X)}(h_e^2) = \mu',$$

since μ', ν and $f^{(X)}(\chi)$ are all $\{0, 1\}$-valued. Thus $\mathbf{D} \models \mathrm{Edg}(\mu)$.

Conversely, suppose that $\mathbf{D} \models \mathrm{Edg}(\mu)$. Choose μ', ν such that

$$\mathbf{D} \models \mathrm{Gen}(\mu') \wedge \mathrm{Gen}(\nu) \wedge \mathrm{Eq}(\mu, \mu')$$

and

$$\nu + f^{(X)}(\chi) = \mu' \text{ or } \mu' + f^{(X)}(\chi) = \nu.$$

Then since $\langle U, +|_U \rangle$ is an Abelian group and $f(\chi(x)) = 0$ for all $x \neq p_1$, while $f(\chi(p_1)) = 1$, we have

$$[\![\nu \neq \mu']\!] = \{p_1\}.$$

By Claim 1, there are $\alpha, \beta \in G^* \cup E^*$ such that $\mathrm{sh}(\nu) = \mathrm{sh}(\alpha)$ and $\mathrm{sh}(\mu') = \mathrm{sh}(\beta)$. The displayed equation above then implies that α and β restricted to $X - \{p_1\}$ have the same shape, and that $\alpha \neq \beta$. (If $\mathrm{sh}(\nu) = \mathrm{sh}(\mu')$, then since $\alpha(p_1) \in \alpha(X - \{p_1\})$ we would have some $x \in X - \{p_1\}$ with $\nu(p_1) = \nu(x) = \mu'(x) = \mu'(p_1)$, because $\mathrm{sh}(\nu) = \mathrm{sh}(\mu')$; but in fact, $\nu(p_1) \neq \mu'(p_1)$.) Now examining the definition of the generators, we see that the only possibility is $\alpha = h_e^i$, $\beta = h_e^j$ for some $e \in E$ and $i \neq j$. So we have $\mathrm{sh}(\mu) = \mathrm{sh}(\mu') = \mathrm{sh}(h_e^j)$ and $\mathbf{D} \models \mathrm{Eq}(\mu, f^{(X)}(h_e^j))$ as asserted (by Claim 1).

Letting $\mathrm{Un}(x)$ be the formula $\mathrm{Gen}(x) \wedge \neg\, \mathrm{Edg}(x)$, we see that $\mathbf{D} \models \mathrm{Un}(\mu)$ iff $\mu \in D(U)$ and $\mathbf{D} \models \mathrm{Eq}(\mu, f^{(X)}(h_v))$ for some $v \in G$. Let $\mathrm{E}(x, y)$ be the formula

$$\begin{aligned} &\mathrm{Un}(x) \wedge \mathrm{Un}(y) \wedge \neg\, \mathrm{Eq}(x, y) \wedge \\ &\quad \exists x'y' \Big(\mathrm{Eq}(x, x') \wedge \mathrm{Eq}(y, y') \wedge \mathrm{Edg}(x' + y') \Big). \end{aligned}$$

Claim 3. For μ, $\nu \in D$ with $\mathbf{D} \models \mathrm{Un}(\mu) \wedge \mathrm{Un}(\nu)$,

$$\mathbf{D} \models \mathrm{E}(\mu, \nu) \text{ if and only if } \mathbf{D} \models \mathrm{Eq}(\mu, f^{(X)}(h_v)) \;\wedge\; \mathrm{Eq}(\nu, f^{(X)}(h_w))$$

for some $v, w \in G$ with $\{v, w\} \in E$.

To prove the claim, let $\mathbf{D} \models \mathrm{E}(\mu, \nu)$, say v, $w \in G$ are such that $\mathrm{Eq}(\mu, f^{(X)}(h_v))$ and $\mathrm{Eq}(\nu, f^{(X)}(h_w))$. Then $v \neq w$ since $\neg\, \mathrm{Eq}(\mu, \nu)$. Choose μ', ν' that witness $\mathrm{E}(\mu, \nu)$. Then $\mathrm{Eq}(\mu' + \nu', f^{(X)}(h_e^i))$ for some $e \in E$ and $i \leq 2$. We'll show that $\{v, w\} = e$.

Since $v \neq w$ and $\mathrm{sh}(\nu') = \mathrm{sh}(\nu) = \mathrm{sh}(h_w)$ then $\nu'(v) = \nu'(p_2)$. For like reasons, $\mu'(v) \neq \mu'(p_2)$. Therefore

$$(\mu' + \nu')(v) \neq (\mu' + \nu')(p_2),$$

implying that

$$h_e^i(v) \neq h_e^i(p_2)$$

since $\mathrm{sh}(\mu' + \nu') = \mathrm{sh}(h_e^i)$. This means that $v \in e$. The same argument shows that $w \in e$, giving $e = \{v, w\}$ since $v \neq w$.

The converse in Claim 3 is easily proved.

We now have shown that the structure

$$\langle \mathrm{Un}^{\mathbf{D}}, \mathrm{E}^{\mathbf{D}} \rangle / \mathrm{Eq}^{\mathbf{D}}$$

is isomorphic to our graph $\mathbf{G} = \langle G, E \rangle$ via the map which sends the $\mathrm{Eq}^{\mathbf{D}}$ class of $f^{(X)}(h_v)$ in $\mathrm{Un}^{\mathbf{D}}$ to v in G. The first order formulas $\mathrm{Un}(x)$, $\mathrm{E}(x, y)$ and $\mathrm{Eq}(x, y)$ are independent of the choice of the graph $\mathbf{G}$, and so we conclude that the class $\boldsymbol{P}_{\mathbf{s}}(\mathbf{A})$ is unstructured. □

The remaining case in the proof of Lemma 8.6 is when the hypotheses of Lemma 10.2 cannot be satisfied. In this case, it follows from Theorem 0.25 that for all $g \in \mathrm{Pol}_1 \mathbf{A}$, if $g(\beta|_V) \subseteq \alpha$ and $g(\alpha|_V) \subseteq 0_A$ then $g(\beta|_V) \subseteq 0_A$. A lemma similar to Lemma 10.1 is needed.

LEMMA 10.3 *Let $t(x_1, \ldots, x_k)$ be a k-ary polynomial operation of* $\mathbf{A}$ *such that $t(A^k) \subseteq V$ and for all $\bar{a} \in A^k$ and all $i \leq k$,*

$$t(a_1, \ldots, a_{i-1}, \alpha|_V, a_{i+1}, \ldots, a_k) \not\subseteq 0_A.$$

If $k > 1$ then

$$t(a_1, \ldots, a_{i-1}, \beta|_V, a_{i+1}, \ldots, a_k) \subseteq \alpha$$

for all $\bar{a} \in A^k$ and $i \leq k$.

PROOF. If this lemma is false then we can find $t(x, y) \in \mathrm{Pol}_2\mathbf{A}$ such that $t(A \times A) \subseteq V$, $t(\alpha|_V, c) \not\subseteq 0_A$ and $t(c, \alpha|_V) \not\subseteq 0_A$ for all $c \in A$ and for some $a \in A$, $t(\beta|_V, a) \not\subseteq \alpha$. Since $\mathrm{typ}(\alpha, \beta) = \mathbf{1}$, $t(A \times A) \subseteq V$, and $t(\beta|_V, a) \not\subseteq \alpha$, then $t(c, \beta|_V) \subseteq \alpha$ for all $c \in A$. This follows from Lemma 0.40 and the fact that $\mathbf{A}/\alpha$ is Abelian.

Now we can find u, $v \in V$ and $d \in A$ such that

$$\langle u, v \rangle \in \alpha - 0_A \quad \text{and} \quad t(d, u) \neq t(d, v).$$

We may choose u and v so that $\{u, v\} \subseteq M' \subseteq U \subseteq V$ for some $U \in M_{\mathbf{A}}(0_A, \alpha)$ and $\langle 0_A, \alpha \rangle$-trace M'. Since $\mathbf{A}$ is Abelian, then $t(u, u) \neq t(u, v)$; letting $0 = t(u, u)$ and $1 = t(u, v)$, we have that $\{0, 1\} \subseteq V$ and $\langle 0, 1 \rangle \in \alpha - 0_A$.

Since $\mathbf{A}/\alpha$ is Abelian and $t(\beta|_V, a) \not\subseteq \alpha$, then $t(\beta|_V, w) \not\subseteq \alpha$ for all $w \in A$. Since the range of t is contained in V and $V \in M_{\mathbf{A}}(\alpha, \beta)$ it follows that $t(V, w) = V$ for all $w \in A$. Thus for some $c \in V$, $t(c, v) = 0 = t(u, u)$. Since $\langle t(c, v), t(u, v) \rangle \in \alpha$ and $t(x, v)$ permutes the $\alpha|_V$ classes, then we must have $\langle c, u \rangle \in \alpha$. It is clear that $c \neq u$ since $0 \neq 1$; and so we can find $h \in \mathrm{Pol}_1\mathbf{A}$ such that

$$h(A) \subseteq U, \quad \{h(u), h(c)\} \subseteq M' \quad \text{and} \quad h(u) \neq h(c).$$

As we have seen in previous arguments, we may choose h so that $h(u) = u$ and $h(c) = v$.

Let g be the unary polynomial operation defined by $g(x) = t(x, h(x))$. Then $g(A) \subseteq V$. Choose r, $s \in V$ with $\langle r, s \rangle \in \beta - \alpha$ and consider $g(r)$ and $g(s)$. We have

$$\langle t(s, h(s)), t(s, h(r)) \rangle \in \alpha$$

since $t(s, \beta|_V) \subseteq \alpha$ and $\langle h(r), h(s) \rangle \in \beta|_V$. But

$$\langle t(r, h(r)), t(s, h(r)) \rangle \in \beta - \alpha$$

since $t(V, h(r)) = V$. Thus $\langle g(r), g(s) \rangle \in \beta - \alpha$. Since r, $s \in V$ and $V \in M_{\mathbf{A}}(\alpha, \beta)$ we conclude that $g(V) = V$. But

$$g(u) = t(u, h(u)) = t(u, u) = 0 = t(c, v) = t(c, h(c)) = g(c),$$

which is impossible since u, $c \in V$ and $u \neq c$. □

LEMMA 10.4 *Let* $\mathbf{A}$, M, N, α, β, f *be as in the statement of Lemma 8.6. Let* $V \in M_{\mathbf{A}}(\alpha,\beta)$ *contain the trace* N, *and suppose that for all* $g \in \mathrm{Pol}_1 \mathbf{A}$, *if* $g(\beta|_V) \subseteq \alpha$ *and* $g(\alpha|_V) \subseteq 0_A$, *then* $g(\beta|_V) \subseteq 0_A$. *Then* $\boldsymbol{P}_s(\mathbf{A})$ *is unstructured.*

PROOF. Since $f(N) \subseteq M$ and f is not constant on N, then $f(\beta|_V) \subseteq \alpha$ and $f(\beta|_V) \not\subseteq 0_A$. Thus by our assumptions, $f(\alpha|_V) \not\subseteq 0_A$. Then it follows easily from Theorem 0.25 that there exist $U, W \in M_{\mathbf{A}}(0_A,\alpha)$ with $U \subseteq V$ and $f(U) = W$. Since $U \simeq W$ in $\mathbf{A}$, without loss of generality we can arrange that $f(A) = f(U) = U$ where $U \in M_{\mathbf{A}}(0_A,\alpha)$, $U \subseteq V$, $f(N) \subseteq M \subseteq U$, and f is nonconstant on N.

Since for $n \in \omega$ the polynomial operation f^n satisfies $f^n(A) = f^n(U) = U$ and f^n is not constant on N, then we may suppose that $f(x) = f^2(x)$ for all $x \in A$ (just take an appropriate power of f).

Our construction to interpret graphs in $\boldsymbol{P}_s(\mathbf{A})$ is a variation of the one we used in proving Lemma 10.2. Choose a, $b \in N$ such that $\langle a,b\rangle \in \beta - \alpha$ and $\langle f(a), f(b)\rangle \in \alpha - 0_A$. (This we can do because f is nonconstant on N and $\beta|_N \neq \alpha|_N$.)

Let $\mathbf{G} = \langle G, E\rangle$ be any graph with $|G| > 2$. Choose three distinct points p_1, p_2, p_3 not in G and let $X = G \cup \{p_1,p_2,p_3\}$. We proceed to define the algebra in which $\mathbf{G}$ can be interpreted. Define $h_v^i : X \to \{a,b\}$, for $v \in G$ and $i \in \{1,2\}$, by

$$h_v^1(x) = \begin{cases} b & \text{if } x = v \text{ or } p_1 \\ a & \text{otherwise} \end{cases}$$

and

$$h_v^2(x) = \begin{cases} b & \text{if } x = v \\ a & \text{otherwise.} \end{cases}$$

Define $h_e^i : X \to \{a,b\}$, for $e \in E$ and $i \in \{1,2\}$, by

$$h_e^1(x) = \begin{cases} b & \text{if } x \in e \text{ or } x = p_2 \\ a & \text{otherwise} \end{cases}$$

and

$$h_e^2(x) = \begin{cases} b & \text{if } x \in e \\ a & \text{otherwise.} \end{cases}$$

Define $\chi_1, \chi_2 : X \to \{a,b\}$ by

$$\chi_i(x) = \begin{cases} b & \text{if } x = p_i \\ a & \text{otherwise.} \end{cases}$$

For $i \in \{1,2\}$ let

$$G_i^* = \{h_v^i : v \in G\},$$
$$E_i^* = \{h_e^i : e \in E\},$$

and let M^* be the set of all nonconstant functions $\mu \in \{a,b\}^X$ such that $\mu(p_1) = \mu(p_2) = \mu(p_3) = a$. We define

D is the subalgebra of $\mathbf{A}^X$ generated by
the set $G_1^* \cup E_1^* \cup M^* \cup \{\chi_1, \chi_2\} \cup \{\hat{a} : a \in A\}$.

Note that **D** is a diagonal subdirect power of **A** and that every $\mu \in D$ is constant modulo β, since this is true of the generators of **D**. Note also that $G_2^* \cup E_2^* \subseteq M^* \subseteq D$.

Recall that $f \in \mathrm{E}(\mathbf{A})$ and $f(A) = U$. Choose $e_2 \in \mathrm{E}(\mathbf{A})$ with $e_2(A) = V$. Now let $\mathrm{U}(x)$, $\mathrm{V}(x)$, and $\mathrm{Con}(x)$ be formulas (which use the parameters $\hat{a}$ $(a \in A)$ and the functions f and e_2) such that $\mathbf{D} \models \mathrm{U}(\mu)$ iff $\mu \in D(U)$, $\mathbf{D} \models \mathrm{V}(\mu)$ iff $\mu \in D(V)$, and $\mathbf{D} \models \mathrm{Con}(\mu)$ iff μ is constant.

Let

$$S = \{p \in \mathrm{Pol}_1\mathbf{A} : p(U) = U\},$$
$$H = \{p \in \mathrm{Pol}_1\mathbf{A} : p(V) = V\},$$

and let $\mathrm{Eq}_U(x,y)$ and $\mathrm{Eq}_V(x,y)$ be formulas such that

$$\mathbf{D} \models \mathrm{Eq}_U(\mu,\nu) \text{ iff } \{\mu,\nu\} \subseteq D(U) \text{ and } \mu = g^{(X)}(\nu) \text{ for some } g \in S$$

while

$$\mathbf{D} \models \mathrm{Eq}_V(\mu,\nu) \text{ iff } \{\mu,\nu\} \subseteq D(V) \text{ and } \mu = h^{(X)}(\nu) \text{ for some } h \in H.$$

As in Lemma 10.2 we have that if μ, $\nu \in D(U)$ are both two-valued then

$$\mathbf{D} \models \mathrm{Eq}_U(\mu,\nu) \text{ iff } \mathrm{sh}(\mu) = \mathrm{sh}(\nu). \tag{10.2}$$

Let us note also that Eq_U and Eq_V define equivalence relations on the respective sets $D(U)$ and $D(V)$.

Claim 1. For all $\mu \in D(V)$, either μ is constant modulo α or $\mu = h^{(X)}(\nu)$ for some nonconstant generator ν and some $h \in H$.

To prove this claim, suppose that $\mu \in D(V)$. We can write $\mu = t^{(X)}(\nu_1, \ldots, \nu_k)$ for some nonconstant generators $\nu_1, \ldots, \nu_k$ and some polynomial operation $t \in \mathrm{Pol}\mathbf{A}$ such that the range of t is contained in V. We may choose t so that for all $\bar{a} \in A^k$ and $i \leq k$,

$$t(a_1, \ldots, a_{i-1}, \alpha|_V, a_{i+1}, \ldots, a_k) \not\subseteq 0_A.$$

For if say, $t(\alpha|_V, a_2, \ldots, a_k) \subseteq 0_A$, then $t(\beta|_V, a_2, \ldots, a_k) \subseteq \alpha$, since $\alpha|_V \neq 0_V$ and $V \in M_{\mathbf{A}}(\alpha, \beta)$. But then by our assumptions, $t(\beta|_V, a_2, \ldots, a_k) \subseteq 0_A$. In this case, we have

$$t^{(X)}(\nu_1, \ldots, \nu_k) = t^{(X)}(\hat{a}, \nu_2, \ldots, \nu_k),$$

and so we could replace the operation $t(x_1, \ldots, x_k)$ by the polynomial operation $t(a, x_2, \ldots, x_k)$.

Now by Lemma 10.3, either

$$t(a_1, \ldots, a_{i-1}, \beta|_V, a_{i+1}, \ldots, a_k) \subseteq \alpha \quad \text{for all } i \leq k \text{ and } \bar{a} \in A^k,$$

in which case we call t a **collapsing function**, or $k = 1$ and $t(V) = V$. If t is a collapsing function then, since $\mathbf{A}$ is Abelian and the ν_i's are constant modulo β and lie in $D(V)$, we have that $t^{(X)}(\nu_1, \ldots, \nu_k)$ is constant modulo α. Thus either μ is constant modulo α or $\mu = h^{(X)}(\nu)$ for some nonconstant generator ν and some $h \in H$. We remark that in the latter case, μ must be nonconstant modulo α, because ν is and h maps V bijectively onto itself. This finishes our proof of Claim 1.

The next claim contains the crux of our argument.

Claim 2. There exists a formula Alpha(x) such that for all $\mu \in D$, $\mathbf{D} \models \text{Alpha}(\mu)$ iff $\mu \in D(V)$ and μ is constant modulo α.

The proof of this claim relies on Lemma 10.5, which follows the proof of this lemma. According to Lemma 10.5, if $t \in \operatorname{Pol}_k \mathbf{A}$ is a collapsing function, and $\bar{\nu}$ is a k-tuple of nonconstant generators, then for some collapsing function $p \in \operatorname{Pol}_j \mathbf{A}$, where $j \leq L = 6|A|^{|A|}$, and for some nonconstant generators $\bar{\lambda}$, we have $t^{(X)}(\bar{\nu}) = p^{(X)}(\bar{\lambda})$. Let I be the set of of all collapsing functions p such that $p \in \operatorname{Pol}_j \mathbf{A}$ for some $j \leq L$ and let

$$\text{Alpha}(x) \overset{\text{def}}{\longleftrightarrow} \text{V}(x) \wedge \left(\text{Con}(x) \vee \bigvee_{p \in I} \exists \bar{y} \left([\bigwedge_i \text{V}(y_i)] \wedge x \approx p^{(X)}(\bar{y}) \right)\right).$$

Now if $\mathbf{D} \models \text{Alpha}(\mu)$ then

$$\mu \in D(V) \text{ and } \mu = p^{(X)}(\lambda_1, \ldots, \lambda_k)$$

for some $p \in I$ and sequence $\bar{\lambda}$ from $D(V)$. Since the λ_i's are all constant modulo β and p is a collapsing function, then μ must be constant modulo α.

Conversely, if $\mu \in D(V)$ is constant modulo α but not constant, then by the proof of Claim 1, $\mu = t^{(X)}(\bar{\nu})$ for some collapsing function t and some nonconstant generators $\bar{\nu}$. By Lemma 10.5, we may assume that $t \in I$. Thus $\mathbf{D} \models \text{Alpha}(\mu)$.

Let Beta(x) be $\text{V}(x) \wedge \neg \text{Alpha}(x)$. Then for $\mu \in D(V)$,

$$\mathbf{D} \models \text{Beta}(\mu) \text{ iff } \mu = h^{(X)}(\nu)$$

for some nonconstant generator ν and some $h \in H$.

We are now interested in isolating those elements μ of $D(U)$ such that $\text{sh}(\mu) = \text{sh}(\nu)$ for some nonconstant generator ν different from χ_1 and χ_2. Let Gen(x) be

$$\text{U}(x) \wedge \neg \text{Con}(x) \wedge \exists y \Big(\text{Beta}(y) \wedge \\ \neg \text{Eq}_V(y, \chi_1) \wedge \neg \text{Eq}_V(y, \chi_2) \wedge \text{Eq}_U(x, f^{(X)}(y))\Big).$$

The verification that

$$\mathbf{D} \models \text{Gen}(\mu) \text{ iff } \mu \in D(U) \text{ and } \text{sh}(\mu) = \text{sh}(\nu)$$

for some nonconstant generator ν different from χ_1 and χ_2 is left to the reader. It is fairly straightforward using (10.2), Claim 1, and the properties of Eq_U and Eq_V. It follows that Gen(μ) if and only if $\mu = k^{(X)} f^{(X)}(\nu)$ for some $k \in S$ and ν a nonconstant generator different from χ_1 and χ_2.

The detail that distinguishes elements of

$$G_1^* \cup G_2^* \cup E_1^* \cup E_2^*$$

from elements of M^* not in the above union is that for μ in the above union there is another nonconstant generator μ' such that $\mu(x) = \mu'(x)$ for all $x \in G$, and $\mu \neq \mu'$. We will now present two formulas that express this fact.

Choose a ternary polynomial operation $d(x, y, z)$ of $\mathbf{A}$ such that $d(A^3) \subseteq U$ and $d|_U$ is a Maltsev operation on U. Let $0 = f(a)$, $1 = f(b)$ in the trace M and let $x + y = d(x, 0, y)$. Since $A|_U$ is Abelian the algebra $\langle U, +|_U\rangle$ is an Abelian group. For $i \in \{1, 2\}$ let $\text{R}_i(x, y)$ be the formula

$$\text{Gen}(x) \wedge \text{Gen}(y) \wedge \neg \text{Eq}_U(x, y) \wedge \exists x', y' \Big(\text{Eq}_U(x, x') \wedge \\ \text{Eq}_U(y, y') \wedge \big[(x' + f^{(X)}(\chi_i) \approx y') \vee (y' + f^{(X)}(\chi_i) \approx x')\big]\Big).$$

Claim 3. For μ, $\nu \in D(U)$ we have

(i) $\mathbf{D} \models \mathrm{R}_1(\mu, \nu)$ if and only if $\mathrm{sh}(\mu) = \mathrm{sh}(h_v^i)$ and $\mathrm{sh}(\nu) = \mathrm{sh}(h_v^j)$ for some $v \in G$ and $\{i, j\} = \{1, 2\}$.

(ii) $\mathbf{D} \models \mathrm{R}_2(\mu, \nu)$ if and only if $\mathrm{sh}(\mu) = \mathrm{sh}(h_e^i)$ and $\mathrm{sh}(\nu) = \mathrm{sh}(h_e^j)$ for some $e \in E$ and $\{i, j\} = \{1, 2\}$.

To prove (i), we note first that if $\mathrm{sh}(\mu) = \mathrm{sh}(h_v^1)$ and $\mathrm{sh}(\nu) = \mathrm{sh}(h_v^2)$ for some $v \in G$ then $f^{(X)}(h_v^1)$ and $f^{(X)}(h_v^2)$ are witnesses for the satisfaction of $\mathrm{R}_1(\mu, \nu)$ and $\mathrm{R}_1(\nu, \mu)$. Conversely, let $\mathbf{D} \models \mathrm{R}_1(\mu, \nu)$. Then $\mathrm{sh}(\mu) = \mathrm{sh}(\lambda_1)$ and $\mathrm{sh}(\nu) = \mathrm{sh}(\lambda_2)$ for some distinct nonconstant generators λ_i different from χ_1 and χ_2. Choose μ', ν' that witness $\mathrm{R}_1(\mu, \nu)$. Then either

$$\nu' + f^{(X)}(\chi_1) = \mu' \quad \text{or} \quad \mu' + f^{(X)}(\chi_1) = \nu'.$$

Without loss of generality, assume the latter. We will show that $\lambda_1(x) = \lambda_2(x)$ for all $x \in X - \{p_1\}$.

Since $f(a) = 0$ and $\chi_1(x) = a$ for all $x \neq p_1$, then $\mu'(x) = \nu'(x)$ for all $x \neq p_1$. But then since $\mathrm{sh}(\mu') = \mathrm{sh}(\lambda_1)$ and $\mathrm{sh}(\nu') = \mathrm{sh}(\lambda_2)$, it follows that λ_1 and λ_2 have the same shape on the set $X - \{p_1\}$. Now since $\lambda_1 \neq \lambda_2$ and $|G| > 2$, it follows upon inspecting the description of the generators that $\{\lambda_1, \lambda_2\} = \{h_v^1, h_v^2\}$ for some $v \in G$.

The proof of the second assertion of Claim 3 is handled similarly.

Now we put

$$\mathrm{Un}(x) \stackrel{\mathrm{def}}{\longleftrightarrow} \exists y\, \mathrm{R}_1(x, y)$$

and

$$\mathrm{Edg}(x) \stackrel{\mathrm{def}}{\longleftrightarrow} \exists y\, \mathrm{R}_2(x, y).$$

Then it should be obvious that $\mathbf{D} \models \mathrm{Un}(\mu)$ if and only if $\mu \in D(U)$ and $\mathrm{sh}(\mu) = \mathrm{sh}(h_v^i)$ for some $v \in G$ and $i \in \{1, 2\}$; and that $\mathbf{D} \models \mathrm{Edg}(\mu)$ if and only if $\mu \in D(U)$ and $\mathrm{sh}(\mu) = \mathrm{sh}(h_e^i)$ for some $e \in E$ and some $i \in \{1, 2\}$. We put

$$\mathrm{Eq}(x, y) \stackrel{\mathrm{def}}{\longleftrightarrow} \mathrm{Un}(x) \wedge \mathrm{Un}(y) \wedge (\mathrm{Eq}_U(x, y) \vee \mathrm{R}_1(x, y)).$$

Thus Eq defines an equivalence relation on $\mathrm{Un}^{\mathbf{D}}$ such that

$$\mathbf{D} \models \mathrm{Eq}(\mu, \nu) \text{ iff } \mathrm{sh}(\mu) = \mathrm{sh}(h_v^i) \text{ and } \mathrm{sh}(\nu) = \mathrm{sh}(h_v^j)$$

for some $v \in G$ and i, $j \in \{1, 2\}$. Finally, put

$$\begin{aligned} \mathrm{E}(x, y) \stackrel{\mathrm{def}}{\longleftrightarrow} & \ \mathrm{Un}(x) \wedge \mathrm{Un}(y) \wedge \neg\, \mathrm{Eq}(x, y) \wedge \\ & \exists x', y' \Big(\mathrm{Eq}(x, x') \wedge \mathrm{Eq}(y, y') \wedge \mathrm{Edg}(x' + y') \Big). \end{aligned}$$

Claim 4. $\mathbf{D} \models \mathrm{E}(\mu, \nu)$ if and only if

$$\mathrm{Un}(\mu),\ \mathrm{Un}(\nu),\ \mathrm{sh}(\mu) = \mathrm{sh}(h_v^i) \text{ and } \mathrm{sh}(\nu) = \mathrm{sh}(h_w^j)$$

for some $i, j \in \{1, 2\}$ with $\{v, w\} \in E$.

To prove this, suppose that $\mathbf{D} \models \mathrm{E}(\mu, \nu)$, and say

$$\mathrm{sh}(\mu) = \mathrm{sh}(h_v^i),\ \mathrm{sh}(\nu) = \mathrm{sh}(h_w^j).$$

Choose μ', ν' such that

$$\mathrm{Eq}(\mu, \mu'),\ \mathrm{Eq}(\nu, \nu') \text{ and } \mathrm{Edg}(\mu' + \nu').$$

Then

$$\mathrm{sh}(\mu') = \mathrm{sh}(h_v^k) \quad \text{and} \quad \mathrm{sh}(\nu') = \mathrm{sh}(h_w^l)$$

for some $k, l \in \{1, 2\}$. Let $e \in E$ and $m \in \{1, 2\}$ be such that $\mathrm{sh}(\mu' + \nu') = \mathrm{sh}(h_e^m)$. Then $\mu' + \nu' = h^{(X)} f^{(X)}(h_e^m)$ for some $h \in S$.

Supposing that $v \notin e$ then evaluating at the coordinate v, we find that

$$\mu'(v) + \nu'(v) = hf(h_e^m(v)) = hf(h_e^m(p_3)).$$

Now evaluating at p_3, we get

$$\mu'(p_3) + \nu'(p_3) = hf(h_e^m(p_3))$$

and so

$$\mu'(p_3) + \nu'(v) = hf(h_e^m(p_3)),$$

since $v \neq w$ implies $\nu'(v) = \nu'(p_3)$. Thus

$$\mu'(v) + \nu'(v) = \mu'(p_3) + \nu'(v)$$

and so $\mu'(v) = \mu'(p_3)$, since $\langle U, +|_U \rangle$ is an Abelian group. This is impossible since $\mathrm{sh}(\mu') = \mathrm{sh}(h_v^k)$, and so we must have $v \in e$. By symmetry, we also conclude that $w \in e$ and so $\{v, w\} = e \in E$.

The converse is fairly easy to prove, and we leave the details to the reader.

Summing up, we have shown that the structure

$$\langle \mathrm{Un}^{\mathbf{D}}, \mathrm{E}^{\mathbf{D}} \rangle / \mathrm{Eq}^{\mathbf{D}}$$

is isomorphic to the graph $\mathbf{G} = \langle G, E \rangle$. As in Lemma 10.3, we conclude that $\boldsymbol{P}_s(\mathbf{A})$ is unstructured. $\square$

LEMMA 10.5 *Let* **A** *and* **D** *be as in Lemma 10.4 and let* $L = 6|A|^{|A|}$. *If* $t(x_1, \ldots, x_m) \in \operatorname{Pol} \mathbf{A}$ *is a collapsing function and* $\nu_1, \ldots, \nu_m$ *are nonconstant generators of* **D**, *then for some collapsing function* $p(x_1, \ldots, x_j) \in \operatorname{Pol}_j \mathbf{A}$ *with* $j \leq L$, *and some nonconstant generators* $\lambda_1, \ldots, \lambda_j$ *we have*

$$\mu = t^{(X)}(\nu_1, \ldots, \nu_m) = p^{(X)}(\lambda_1, \ldots, \lambda_j).$$

PROOF. This proof makes heavy use of the Abelian property of **A**. The main ideas for the proof come from J. Berman and R. McKenzie [1984]. Let $[m] = \{1, \ldots, m\}$ and for $i, j \in [m]$, say that t is $\langle i, j\rangle$-symmetric if the value of t is never changed by exchanging the i-th and j-th variables of t. For $I \subseteq [m]$, say that t is I-symmetric iff t is $\langle i, j\rangle$-symmetric for all $\{i, j\} \subseteq I$. Call t totally symmetric iff t is $[m]$-symmetric. As in Section 4 of Berman and McKenzie, we define the equivalence relation θ_t on $[m]$ by specifying that $\langle i, j\rangle \in \theta_t$ iff t is $\langle i, j\rangle$-symmetric. Thus t is I-symmetric for all θ_t blocks I. Let $l = |A|$. Then Lemma 4.6 of Berman and McKenzie tells us that the equivalence relation θ_t has at most $2l^{l-1}$ blocks.

We first assume that our function t is totally symmetric and $m \geq 3l$. Consider the sequence

$$t(a, a, \ldots, a),\ t(b, a, \ldots, a), \ldots, t(b, b, b, \ldots, b, a),\ t(b, b, \ldots, b).$$

Since $m \geq 3l$, then there is a repetition in this sequence. So by symmetry and the Abelian property, $t(b, b, \ldots, b, a, \ldots, a) = t(a, a, \ldots, a)$ for a certain non-zero number n of b's. Let n be the least positive number such that the above equality holds. Note that $n \leq l$. Again using symmetry, we can conclude that if $\bar{z}$ is an m-tuple of a's and b's, then we can find a sequence $\bar{z}' = \langle b, b, \ldots, b, a, \ldots, a\rangle$ with less than n b's such that $t(\bar{z}) = t(\bar{z}')$.

Now let $\nu_1, \ldots, \nu_m$ be nonconstant generators of **D** and let

$$\mu = t^{(X)}(\nu_1, \ldots, \nu_m).$$

It follows from the above discussion that we can find numbers p, k_1, $k_2 < n$ and generators $\lambda_1, \ldots, \lambda_p \in M^*$ such that

$$\mu = t^{(X)}(\lambda_1, \ldots, \lambda_p, \chi_1, \ldots, \chi_1, \chi_2, \ldots, \chi_2, \hat{a}, \ldots, \hat{a}),$$

where χ_i occurs k_i times.

For the general case, let $I_1, \ldots, I_s$ be the θ_t equivalence classes. Recall that $s \leq 2l^{l-1}$. Using the Abelian property and the above result for totally symmetric functions, for each $i \leq s$, we may replace the set

$$\{\nu_j : j \in I_i\}$$

by a set of at most $3l$ nonconstant generators, along with a suitable number of occurrences of $\hat{a}$, and preserve the value of $\mu = t^{(X)}(\nu_1, \ldots, \nu_m)$. Now since there are at most $2l^{l-1}$ equivalence classes, then we have shown that

$$\mu = t^{(X)}(\lambda_1, \ldots, \lambda_m)$$

where at most $6l^l = L$ of the λ_i's are nonconstant generators, and the rest are equal to $\hat{a}$. Thus by replacing the appropriate variables of t by a, we can construct a polynomial operation $p(x_1, \ldots, x_j)$ of arity $\leq L$ and nonconstant generators $\lambda_1, \ldots, \lambda_j$ such that p is a collapsing function and $\mu = p^{(X)}(\lambda_1, \ldots, \lambda_j)$. This completes our proof of the lemma. □

Chapter 11

From strongly Abelian to essentially unary varieties

In this chapter and the next, we characterize the decidable, strongly Abelian, locally finite varieties. In this chapter, we reduce the task to the seemingly more modest one of determining those locally finite varieties of multi-sorted unary algebras that are decidable. In the next chapter, we solve this more modest problem.

Early on in the proof, we will shift our attention to multi-sorted algebras, as this will make our proof easier to visualize and present.

Definition 11.1 Let k be a positive integer.

(1) A **k-sorted similarity type** is a triple $\mathsf{T} = (k, \Phi, \tau)$, where Φ is a set and τ is a function

$$\tau : \Phi \longrightarrow \bigcup_{n>0} \{1, 2, \ldots, k\}^n.$$

The members of Φ are called the **operation symbols** of T.

(2) A **k-sorted algebra of type T** is a system

$$\mathbf{B} = \langle B_1, \ldots, B_k; f^{\mathbf{B}}(f \in \Phi) \rangle,$$

where the B_j are pairwise disjoint nonvoid sets and for $f \in \Phi$, $f^{\mathbf{B}}$ is a function

$$f^{\mathbf{B}} : B_{i_1} \times \cdots \times B_{i_s} \longrightarrow B_{i_{s+1}},$$

where $\tau(f) = \langle i_1, \ldots, i_s, i_{s+1} \rangle$. The sequence $\tau(f)$ is called the **type** of the operation symbol f (and of the operation $f^{\mathbf{B}}$), and i_{s+1} is called the **sort** of f (and of $f^{\mathbf{B}}$). Elements of B_i are called elements of $\mathbf{B}$ of sort i.

The difference between ordinary (1-sorted) algebras and k-sorted algebras is that the latter have multiple universes, and, in their operations, each variable is restricted to range over a specific universe. A k-sorted algebra of type T,

$$\mathbf{B} = \langle B_1, \ldots, B_k; f^{\mathbf{B}}(f \in \Phi)\rangle,$$

can be converted into a model

$$\mathbf{C} = \langle B_1 \cup \cdots \cup B_k; U_1^{\mathbf{C}}, \ldots, U_k^{\mathbf{C}}, r_f^{\mathbf{C}}(r_f \in \Sigma)\rangle,$$

of a first order language L. In $\mathbf{C}$, the sets B_i are specified by unary relations $U_i^{\mathbf{C}}$, and a function $f^{\mathbf{B}}$ ($f \in \Phi$), with $\tau(f) = \langle i_1, \ldots, i_s, i_{s+1}\rangle$, is replaced by its graph, an $s+1$-ary relation $r_f^{\mathbf{C}}$. Through this device of converting k-sorted algebras of type T into models of an ordinary first order language, we may speak of decidable and undecidable classes of these systems, and also of unstructured and ω-unstructured classes.

However, we prefer to treat k-sorted algebras like ordinary algebras, so far as possible. In fact, we shall need to deal with the term operations of a k-sorted algebra, with terms themselves, and with varieties of k-sorted algebras and free algebras in such varieties. Most of the notions and theorems of the general theory of algebras carry over in a fairly routine manner to the theory of k-sorted algebras. The reader can consult W. Taylor [1973] for any details that are not supplied below.

To form the **terms** of the k-sorted type $\mathsf{T} = (k, \Phi, \tau)$, we introduce k disjoint infinite sets of variables, so that each variable has its sort, which is an integer between 1 and k. Terms are then defined so that each term also has its sort. The set of T-terms is the smallest set of sequences (of variables and operation symbols) such that every variable is a term, and if $t_1, \ldots, t_s$ are any terms of respective sorts $i_1, \ldots, i_s$, and if $f \in \Phi$ has type $\tau(f) = \langle i_1, \ldots, i_s, i_{s+1}\rangle$, then $ft_1 \ldots t_s$ is a term, of sort i_{s+1}. Note that every term has a unique sort.

Now if $\mathbf{B}$ is a k-sorted algebra of type T, then terms, as defined above, give rise to multi-sorted operations. If $t = t(x_1, \ldots, x_n)$ is a term in which occur at most the distinct variables $x_1, \ldots, x_n$, of respective sorts $i_1, \ldots, i_n$, and t is of sort i_{n+1}, then we have a corresponding function

$$t^{\mathbf{B}} : B_{i_1} \times \cdots \times B_{i_n} \longrightarrow B_{i_{n+1}},$$

of type $\mu = \langle i_1, \ldots, i_n, i_{n+1}\rangle$, defined in the usual way. (The proper definition proceeds by induction on the length of t.) We say that $t^{\mathbf{B}}$ is a **term**

operation of $\mathbf{B}$, of type μ and sort i_{n+1}. In this context, we shall also say that t is a term of type μ. Note that the type of a term, unlike its sort, is not uniquely determined.

Let $\mathbf{B}$ be a k-sorted algebra of type T. If $\mu = \langle i_1, \ldots, i_s, i_{s+1} \rangle$ is a possible type of a k-sorted operation, (i.e., if $\{i_1, \ldots, i_{s+1}\} \subseteq \{1, \ldots, k\}$) then $\mathrm{Clo}_\mu \mathbf{B}$ denotes the set of all term operations $t^{\mathbf{B}}$ of $\mathbf{B}$ having type μ. We put

$$\mathrm{Clo}\, \mathbf{B} = \bigcup_\mu \mathrm{Clo}_\mu \mathbf{B},$$

and call this set the clone of term operations of $\mathbf{B}$. The clone of polynomial operations of $\mathbf{B}$, denoted Pol $\mathbf{B}$, is the set of multi-sorted operations that are obtained by substituting constants (of the appropriate sorts) for some of the variables in term operations of $\mathbf{B}$. Note that the set Clo $\mathbf{B}$ is identical with the closure under (multi-sorted) composition of the set consisting of the basic operations $f^{\mathbf{B}}$ $(f \in \Phi)$ of $\mathbf{B}$ and the trivial operations $p(x_1, \ldots, x_n) = x_i$.

We define **Abelian** and **strongly Abelian** k-sorted algebras in the same way we defined these concepts for one-sorted algebras, in terms of certain implications holding for all term operations (see Definitions 0.6, 0.7, and 0.12). We point out that the multi-sorted analog of Theorem 0.17 (ii) is true; and it is often easier to derive consequences of the strongly Abelian property using this theorem. Notice that if a type $\mathsf{T} = (k, \Phi, \tau)$ is such that τ assigns at most one variable to each $f \in \Phi$, i.e., if T is a **unary** k-sorted type, then every algebra of type T is strongly Abelian. More generally, if $\mathbf{B}$ is a k-sorted algebra of some type T, and every operation of $\mathbf{B}$ depends on at most one of its variables, then $\mathbf{B}$ is strongly Abelian.

Let $\mathbf{B}$ be a k-sorted algebra. If $\overline{C} = \langle C_1, \ldots, C_k \rangle$ with $\emptyset \neq C_i \subseteq B_i$ for $1 \leq i \leq k$, then $\mathrm{Sg}^{\mathbf{B}}(\overline{C})$ denotes the **subalgebra of B generated by** $\overline{C}$; it is the smallest subalgebra containing each set C_i. If $\overline{P} = \langle P_1, \ldots, P_k \rangle$, where $P_i \subseteq B_i^2$ for $1 \leq i \leq k$, then $\mathrm{Cg}^{\mathbf{B}}(\overline{P})$ denotes the **congruence on B generated by** $\overline{P}$; it is the smallest system $\overline{E} = \langle E_1, \ldots, E_k \rangle$ of equivalence relations, E_i on B_i, compatible with the operations of $\mathbf{B}$, and satisfying $P_i \subseteq E_i$ for $1 \leq i \leq k$.

An **equation** for k-sorted algebras of type T is just a formal expression $s \approx t$ where s and t are T-terms of the same type. It can be proved that a class of k-sorted algebras of the same type is closed under subalgebras, direct products, and homomorphic images (appropriately defined) if and only if it is defined by a set of equations. A class with these properties will be called a **variety** of k-sorted algebras. For $\mathcal{V}$ a variety of k-sorted algebras, and $\overline{X} = \langle X_1, \ldots, X_k \rangle$ a system of nonvoid disjoint sets, $\mathbf{F}_{\mathcal{V}}(\overline{X})$ will denote the $\mathcal{V}$-free algebra generated by $\overline{X}$. As in the 1-sorted case, the

$\mathcal{V}$-free algebras have the universal mapping property for the class $\mathcal{V}$ over the system $\overline{X}$, and so an equation $t_0 \approx t_1$ is valid in $\mathcal{V}$ iff this equation holds in the appropriate $\mathcal{V}$-free algebra.

For the remainder of this chapter, $\mathcal{V}$ denotes an arbitrary but fixed locally finite, strongly Abelian variety of 1-sorted algebras. By Theorem 0.17, $\mathcal{V}$ is finitely generated. We assume that $\mathbf{A}$ is a finite algebra that generates $\mathcal{V}$. To avoid triviality, we assume that $|A| > 1$. Again by Theorem 0.17, no term operation of $\mathbf{A}$ depends on more than $|A|$ of its variables. Our first goal is to show that $\mathcal{V}$ is definitionally equivalent, in a natural way, with a variety $\mathcal{V}[d]$ of k-sorted algebras for a certain $k \leq |A|$. After accomplishing that, we shall show that unless $\mathcal{V}$ is both unstructured and ω-unstructured, then $\mathcal{V}[d]$ consists of essentially unary k-sorted algebras.

Definition 11.2 If $f : A^n \to A$, we say that f is **δ-injective** iff the function $f_\delta(x) = f(x, \ldots, x)$ is one-to-one on A.

As we remarked above, no term operation of $\mathbf{A}$ depends on more than $|A|$ of its variables. We define k ($\leq |A|$) to be the maximum number such that $\mathbf{A}$ has a δ-injective term operation that depends on k variables.

LEMMA 11.3 *There is a term $d(x_1, \ldots, x_k)$ such that $d^{\mathbf{A}}$ is essentially k-ary and the equations*

$$\begin{aligned} d(x, \ldots, x) &\approx x, \\ d(d(x_1^1, \ldots, x_k^1), \ldots, d(x_1^k, \ldots, x_k^k)) &\approx d(x_1^1, x_2^2, \ldots, x_k^k) \end{aligned} \tag{11.1}$$

are valid in $\mathbf{A}$.

PROOF. We choose a term t such that $t^{\mathbf{A}}$ is k-ary, depends on all its variables, and is δ-injective. Then $t_\delta^{\mathbf{A}}$ is a permutation of A, and so we have that

$$\mathbf{A} \models t_\delta^n(x) \approx x$$

for some $n > 0$, since A is finite. Let $d(\bar{x}) = t_\delta^{n-1}(t(\bar{x}))$. Then d is an essentially k-ary, δ-injective term for $\mathbf{A}$, and

$$\mathbf{A} \models d(x, \ldots, x) \approx x.$$

To see that d satisfies the second of equations (11.1), note that if $u_1, \ldots, u_k$ are in A then, where $u = d^{\mathbf{A}}(u_1, \ldots, u_k)$, we have

$$d^{\mathbf{A}}(u, \ldots, u) = d^{\mathbf{A}}(u_1, \ldots, u_k).$$

Then since $\mathbf{A}$ is strongly Abelian, for any $a_1, \ldots, a_k \in A$ and for any i between 1 and k, we have

$$d^{\mathbf{A}}(a_1, \ldots, a_{i-1}, u, a_{i+1}, \ldots, a_k) = d^{\mathbf{A}}(a_1, \ldots, a_{i-1}, u_i, a_{i+1}, \ldots, a_k).$$

The desired equation is an easy consequence of the two equations displayed above. □

Any operation satisfying equations (11.1) is called a k-ary **decomposition operation** . The reader may wish to refer to R. McKenzie, G. McNulty, W. Taylor [1987] for further details of decomposition operations. For the remainder of this chapter, d denotes a fixed term such that $d^{\mathbf{A}}$ is a k-ary decomposition operation on A which depends on all its variables, and $\mathbf{A}$ has no δ-injective term operation depending on more than k variables. It follows that for every $\mathbf{B} \in \mathcal{V}$ $(= \mathbf{V}(\mathbf{A}))$, the operation $d^{\mathbf{B}}$ is a k-ary decomposition operation on the universe of $\mathbf{B}$.

LEMMA 11.4 *For each* $\mathbf{B} \in \mathcal{V}$, *there is an algebra* $\mathbf{B}' \in \mathcal{V}$ *isomorphic to* $\mathbf{B}$ *such that*

$$B' = B_1 \times \cdots \times B_k$$

for some pairwise disjoint nonvoid sets $B_1, \ldots, B_k$, *and*

$$d^{\mathbf{B}'}(\bar{a}^1, \ldots, \bar{a}^k) = \langle a_1^1, \ldots, a_k^k \rangle$$

for all $\bar{a}^1, \ldots, \bar{a}^k \in B'$, *where* $\bar{a}^i = \langle a_1^i, \ldots, a_k^i \rangle$.

PROOF. We begin by choosing some $b \in B$. For $i \leq k$, let $\sim_i$ be the equivalence relation on B such that $x \sim_i y$ iff

$$\mathbf{B} \models d(b, \ldots, b, x, b, \ldots, b) = d(b, \ldots, b, y, b, \ldots, b)$$

where x and y occur at the ith place. From equations (11.1), it follows that $\sim_i$ is independent of the choice of b and that $d^{\mathbf{B}}(b_1, \ldots, b_k) \sim_i b_i$. Now for $1 \leq i \leq k$, let $B_i = B/\sim_i$ and let

$$B' = B_1 \times \cdots \times B_k.$$

Let $\phi : B \rightarrow B'$ be the mapping $x \mapsto \langle x/\sim_1, \ldots, x/\sim_k \rangle$. Using the equations (11.1), it is now easy to see that ϕ is a bijection of B onto B'. Thus ϕ is an isomorphism of $\mathbf{B}$ with a certain uniquely determined algebra $\mathbf{B}'$ with universe B'. From the fact already noticed, that $d^{\mathbf{B}}(b_1, \ldots, b_k) \sim_i b_i$, it follows that $d^{\mathbf{B}'}$ is the operation specified by this lemma. If the sets $B_1, \ldots, B_k$ are not pairwise disjoint, they can be replaced by a sequence of sets in bijective correspondence with them, that are. □

The operation $d^{\mathbf{B}'}$ specified in the lemma is called the **diagonal operation**, or the **standard decomposition operation**, on the product set. Without loss of generality, we can assume, for the remainder of this chapter, that $A = A_1 \times \cdots \times A_k$ and that $d^{\mathbf{A}}$ is the diagonal operation on A. Notice that each of the sets A_i has at least two elements, since $d^{\mathbf{A}}$ depends on all of its variables.

Now observe that any n-ary operation on A is uniquely determined by the sequence of k multi-sorted functions $\langle f_1, \ldots, f_k \rangle$, with

$$f_i : A_1^n \times \cdots \times A_k^n \longrightarrow A_i,$$

that are specified as follows: if $\bar{a}^i = \langle a_1^i, \ldots, a_k^i \rangle \in A$, for $1 \leq i \leq n$, and if

$$\bar{b} = \langle a_1^1, \ldots, a_1^n, \ldots, a_k^1, \ldots, a_k^n \rangle,$$

then

$$f(\bar{a}^1, \ldots, \bar{a}^n) = \langle f_1(\bar{b}), \ldots, f_k(\bar{b}) \rangle.$$

We call f_i the i th **projection** of f and write $p_i f$ for f_i.

We are going to define a k-sorted algebra $\mathbf{A}[d]$ that is intimately associated with $\mathbf{A}$. First we define the type T of this multi-sorted algebra. We can suppose that $\mathbf{A}$ is a model for the language $\mathsf{L} = (\Phi, \rho)$. We put $\mathsf{T} = (k, \Phi', \tau)$, where $\Phi' = \Phi \times \{1, \ldots, k\}$, and for $\langle f, j \rangle \in \Phi'$ we put

$$\tau(\langle f, j \rangle) = \langle 1, \ldots, 1, 2, \ldots, 2, \ldots, k, \ldots, k, j \rangle,$$

consisting of n occurrences each of 1, 2, ..., k and a final j, where $n = \rho(f)$. Now we define

$$\mathbf{A}[d] = \langle A_1, \ldots, A_k; p_j f^{\mathbf{A}} (\langle f, j \rangle \in \Phi') \rangle,$$

a k-sorted algebra of type T.

This same construction can be applied to any algebra $\mathbf{B} \in \mathcal{V}$ to get a multi-sorted algebra $\mathbf{B}[d]$. (First form $\mathbf{B}'$ as in Lemma 11.4, and then form $\mathbf{B}'[d]$.) We define

$$\mathcal{V}[d] = \{\mathbf{C} : \mathbf{C} \cong \mathbf{B}[d] \text{ for some } \mathbf{B} \in \mathcal{V}\}.$$

It is easy to see that $\mathbf{B} \cong \mathbf{B}' \leftrightarrow \mathbf{B}[d] \cong \mathbf{B}'[d]$ when $\mathbf{B}$, $\mathbf{B}' \in \mathcal{V}$; in fact, a natural construction converts $\mathbf{B}[d]$ into an isomorphic copy of $\mathbf{B}$.

Remark 11.5 As we shall note in the first lemma below, $\mathcal{V}[d]$ is a strongly Abelian variety of k-sorted algebras. In fact, the equations that define $\mathcal{V}$ can be converted into equations of type T (each L-equation becomes a set of k T-equations), and the resulting equations define $\mathcal{V}[d]$. The two varieties, $\mathcal{V}$ and $\mathcal{V}[d]$, are equivalent in a very strong sense—in particular, $\mathcal{V}$ is decidable, hereditarily undecidable, unstructured, or ω-unstructured if and only if $\mathcal{V}[d]$ has the same respective property. These assertions are formalized in the first lemma below, but the details of their verification are left to the reader.

Remark 11.6 It may be instructive to observe that every k-sorted algebra of any type is equivalent with (has the same clone of term operations as) some algebra of the form $\mathbf{B}[e]$ where $\mathbf{B}$ is a 1-sorted algebra and e is a term in its language that defines a k-ary decomposition operation on B. In fact, every variety of k-sorted algebras is definitionally equivalent with some variety of the form $\mathcal{W}[e]$. One avenue to developing the theory of varieties of k-sorted algebras is to pass back and forth between $\mathcal{W}$ and $\mathcal{W}[e]$.

Definition 11.7 Let $f(x_1, \ldots, x_n)$ and $g(y_1, \ldots, y_m)$ be two multi-sorted operations over a k-sorted domain $\langle X_1, \ldots, X_k \rangle$. We say that g is a **specialization** of f if there are variables $z_1, \ldots, z_n$ such that z_i has the same sort as x_i for $1 \leq i \leq n$, $\{z_1, \ldots, z_n\} \supseteq \{y_1, \ldots, y_m\}$, and the equation

$$g(y_1, \ldots, y_m) \approx f(z_1, \ldots, z_n)$$

holds. We extend this definition to terms in the natural way.

LEMMA 11.8

(i) *If* $\mathbf{B} \in \mathcal{V}$ *and* $n < \omega$, *then* $\mathrm{Clo}_n\mathbf{B}$ *consists of all the* n*-ary operations* f *over* B *such that for* $1 \leq i \leq k$, *the function* $p_i f$ *is in* $\mathrm{Clo}\,\mathbf{B}[d]$.

(ii) *If* $\mathbf{B} \in \mathcal{V}$, *then every* $g \in \mathrm{Clo}\,\mathbf{B}[d]$ *is equal to a specialization of* $p_i f$ *for some* $f \in \mathrm{Clo}\,\mathbf{B}$ *and some* $1 \leq i \leq k$.

(iii) $\mathbf{A}$ *is essentially* k*-ary if and only if* $\mathbf{A}[d]$ *is essentially unary.*

(iv) $\mathcal{V}[d]$ *is a* k*-sorted strongly Abelian variety.*

(v) *The mapping* $\mathbf{B} \mapsto \mathbf{B}[d]$ *constitutes an equivalence between* $\mathcal{V}$ *and* $\mathcal{V}[d]$ *(regarded as algebraic categories).*

(vi) $\mathcal{V}$ *is undecidable, hereditarily undecidable,* ω*-unstructured or unstructured iff* $\mathcal{V}[d]$ *has the same respective property.*

PROOF. Statements (i) and (ii) can be proved by a straightforward induction, using the characterization of these clones as the smallest set of operations including the basic operations and trivial operations, and closed under composition. The use of the term d is essential in these proofs.

For (iii), suppose that $f(x_1, \ldots, x_{k+1})$ is an essentially $k+1$-ary member of Clo $\mathbf{A}$. Using the pigeonhole principle, we conclude that for some $i \leq k$, $p_i f$ must depend on at least two variables. Conversely, suppose that $\mathbf{A}[d]$ is not essentially unary, say $g \in \operatorname{Clo} \mathbf{A}[d]$ depends on at least two variables. Then by (ii) we can find $f \in \operatorname{Clo}_n \mathbf{A}$ (for some n) such that $p_i f$ is not essentially unary for some i. The operation

$$d^{\mathbf{A}}(y_1, \ldots, y_{i-1}, f(d^{\mathbf{A}}(\bar{x}_1), \ldots, d^{\mathbf{A}}(\bar{x}_n)), y_{i+1}, \ldots, y_k),$$

where all of the above variables are distinct, will depend on at least $k+1$ variables.

To prove that $\mathcal{V}[d]$ is a variety, and to prove (v), it suffices to show that the mapping $\mathbf{B} \mapsto \mathbf{B}[d]$ commutes with the formation of subalgebras, products, and homomorphic images, and that there are bijections

$$\pi_{\mathbf{B}_0, \mathbf{B}_1} : \hom(\mathbf{B}_0[d], \mathbf{B}_1[d]) \leftrightarrow \hom(\mathbf{B}_0, \mathbf{B}_1)$$

which respect the composition of homomorphisms. These facts are easy to establish, and so we will not furnish the details.

That $\mathcal{V}[d]$ is strongly Abelian follows easily from (ii), using the term d.

Lemma 11.4 provides us with a bi-interpretation between $\mathcal{V}$ and $\mathcal{V}[d]$, using the term d and the definable equivalence relations $\sim_1$ through $\sim_k$. Statement (vi) follows from this. □

As we mentioned earlier, our goal in this chapter is to prove the following theorem.

THEOREM 11.9 *Either $\mathcal{V}$ is both unstructured and ω-unstructured, or $\mathcal{V}[d]$ is essentially unary (and $\mathcal{V}$ is essentially k-ary).*

In order to prove this theorem, we assume henceforth that $\mathcal{V}[d]$ is not essentially unary. We will interpret the class of bi-partite graphs into $\mathcal{V}[d]$ in such a way that finite graphs get interpreted into finite algebras. From this it will follow that $\mathcal{V}[d]$ is both unstructured and ω-unstructured. Using Lemma 11.8 (vi) we will then be able to conclude that $\mathcal{V}$ itself is unstructured and ω-unstructured. In what follows, "term" means always term of the type of the algebra $\mathbf{A}[d]$. When writing a term as $t(x_1, \ldots, x_n)$, we always assume, of course, that $x_1, \ldots, x_n$ are distinct variables. The sort of a variable x, or term t, or element b of a k-sorted algebra, will be denoted by $\sigma(x)$, $\sigma(t)$, or $\sigma(b)$, respectively.

Definition 11.10

(1) Let $t(x_1, \ldots x_n)$ be a term, let $1 \leq j \leq n$, and suppose that $\sigma(x_j) = u$. We will say that t is **left invertible** at the variable x_j if there is a term $s(y_1, \ldots, y_m)$ of sort u such that $\sigma(t) = \sigma(y_1)$, the variables $y_1, \ldots, y_m$ are distinct and different from the x's, and

$$\mathbf{A}[d] \models s(t(\bar{x}), y_2, \ldots, y_m) \approx x_j.$$

(2) A term $s(x_1, \ldots, x_n)$ of sort u will be called **right invertible** if there are terms $s_j(y, \bar{z})$ with $\sigma(s_j) = \sigma(x_j)$ for $1 \leq j \leq n$, such that $\sigma(y) = u$ and

$$\mathbf{A}[d] \models s(s_1(y, \bar{z}), \ldots, s_n(y, \bar{z})) \approx y.$$

(3) A term $s(x_1, \ldots, x_s)$ will be called **essentially unary** if the function $s^{\mathbf{A}[d]}$ depends on at most one of its variables.

LEMMA 11.11 *Let $t(x_1, x_2, \ldots, x_s)$ be a term (in the type of $\mathbf{A}[d]$).*

(i) *If $\sigma(t) = \sigma(x_1) = \sigma(x_2)$ and*

$$\mathbf{A}[d] \models t(x, x, \bar{y}) \approx x,$$

then the term t is essentially unary.

(ii) *If the term t is right invertible, then it is essentially unary.*

PROOF. For (i), if $\mathbf{A}[d] \models t(x, x, \bar{y}) \approx x$, then by the strongly Abelian property, t depends on at most the variables x_1 and x_2. By Lemma 11.8 we may assume that $t^{\mathbf{A}[d]}$ is a specialization of the operation $p_i g$ for some $g \in \mathrm{Clo}_n \mathbf{A}$ and some $n > 0$ and $i \leq k$. It follows that the term operation

$$d^{\mathbf{A}}(x_1, \ldots, x_{i-1}, g(z_1, \ldots, z_n), x_{i+1}, \ldots, x_k)$$

is a δ-injective member of Clo $\mathbf{A}$ and so must have essential arity at most k. Now if t depends on both x_1 and x_2, the above term operation would have essential arity greater than k, and so we conclude that t is essentially unary.

For (ii), suppose that t is right invertible, say for certain terms $s_i(y, \bar{z})$ we have

$$\mathbf{A}[d] \models t(s_1(y, \bar{z}), \ldots, s_s(y, \bar{z})) \approx y.$$

Then the term

$$\hat{t}(y_1, \ldots, y_s, \bar{z}) = t(s_1(y_1, \bar{z}), \ldots, s_s(y_s, \bar{z}))$$

must be essentially unary by applying part (i) of this lemma several times, say it depends only on y_1.

Let a, $b_1, \ldots, b_s$, $\bar{c}$ be arbitrary elements of $A[d]$ of the appropriate sort and let

$$u = t^{\mathbf{A}[d]}(b_1, \ldots, b_s).$$

Since u is also equal to

$$t^{\mathbf{A}[d]}(s_1^{\mathbf{A}[d]}(u, \bar{c}), \ldots, s_s^{\mathbf{A}[d]}(u, \bar{c})) = \\ t^{\mathbf{A}[d]}(s_1^{\mathbf{A}[d]}(u, \bar{c}), s_2^{\mathbf{A}[d]}(a, \bar{c}), \ldots, s_s^{\mathbf{A}[d]}(a, \bar{c}))$$

and $\mathbf{A}[d]$ is strongly Abelian it follows that the term t is independent of all but its first variable. □

We have assumed that $\mathbf{A}[d]$ is not essentially unary. Thus there exists a term operation $g(x_1, x_2, \ldots, x_s)$ that depends on both x_1 and x_2. The next lemma asserts that we can choose $g = q^{\mathbf{A}[d]}$ so that q is not left invertible at either of the variables x_1 or x_2.

LEMMA 11.12 *There is a term $q(x_1, \ldots, x_s)$ such that*

(i) *$q^{\mathbf{A}[d]}$ depends on x_1 and x_2.*

(ii) *q is not left invertible at x_1 or at x_2.*

PROOF. We begin by choosing an arbitrary term $t(x_1, x_2, \ldots, x_s)$ that depends on x_1 and x_2 in $\mathbf{A}[d]$. Suppose that t is left invertible at x_1. We can choose a term $s(y, \bar{z})$ such that

$$\mathbf{A}[d] \models s(t(\bar{x}), \bar{z}) \approx x_1.$$

By the strongly Abelian property, s depends only upon y. Let

$$t'(y_1, x_2, \ldots, x_s, \bar{z}) = t(s(y_1, \bar{z}), x_2, \ldots, x_s).$$

Then since $t^{\mathbf{A}[d]}$ depends on x_1 and x_2, and s induces a surjective function of $A_{\sigma(t)}$ onto $A_{\sigma(x_1)}$, we conclude that t' depends on y_1 and x_2 in $\mathbf{A}[d]$.

We now show that t' can't be left invertible at y_1. Consider the term

$$p(x, y, z, \bar{w}) = t'(t'(x, y, \bar{w}), z, \bar{w}).$$

It is not hard to show that if t' is left invertible at y_1 then the term p must depend on the variable y since t' depends on x_2. But by unraveling the definition of p one can show that

$$\mathbf{A}[d] \models p(x, y, z, \bar{w}) \approx t'(x, z, \bar{w}),$$

implying that p is independent of y.

We can use the same argument on the second variable of t' to finally arrive at a term q as desired. □

We introduce a construction with $\mathcal{V}[d]$-free algebras that will be used in the upcoming interpretation. Let $\mathbf{F}$ be the $\mathcal{V}[d]$-free algebra $\mathbf{F}_{\mathcal{V}[d]}(\overline{X})$ where $\overline{X} = \langle X_1, \ldots, X_k \rangle$ and choose some element $\mathbf{z} \notin F$. Put $\overline{X}' = \langle X_1', X_2, \ldots, X_k \rangle$ where $X_1' = X_1 \cup \{\mathbf{z}\}$, and let $\mathbf{F}' = \mathbf{F}_{\mathcal{V}[d]}(\overline{X}')$. Now, choosing an element $0 \in F_1$, we have $\mathbf{F} < \mathbf{F}'$, $0 \in F$, $\mathbf{z} \in F' - F$, and $\sigma(0) = \sigma(\mathbf{z}) = 1$. For $1 \leq i \leq k$, we define

$$C_i = \{\langle t^{\mathbf{F}'}(0, \bar{u}), t^{\mathbf{F}'}(\mathbf{z}, \bar{u}) \rangle : \sigma(t) = i, \ \bar{u} \in F, \text{ and } t(x, \bar{y}) \text{ is not left invertible at } x\}.$$

Then we define a congruence on $\mathbf{F}'$:

$$\theta(\overline{X}, 0, \mathbf{z}) = \mathrm{Cg}^{\mathbf{F}'}(\langle C_1, \ldots, C_k \rangle).$$

LEMMA 11.13 *Let $\theta = \theta(\overline{X}, 0, \mathbf{z})$ and let $a \in F$ with $\sigma(a) = i$. Then*

$$a/\theta = \{b \in F' : \langle a, b \rangle \in C_i\}$$

and

$$\mathbf{z}/\theta = \{\mathbf{z}\}.$$

PROOF. Let $a \in F_i$ and put

$$S = \{b \in F_i' : \langle a, b \rangle \in C_i\}.$$

Clearly, we have $S \subseteq a/\theta$, and $a \in S$. To prove that $a/\theta = S$, it will suffice to prove that if $\langle u, v \rangle \in C_j$ for some $1 \leq j \leq k$ and p is a unary polynomial operation of $\mathbf{F}'$ with $p(u)$ or $p(v)$ in S, then $\{p(u), p(v)\} \subseteq S$.

There are two cases to consider, $p(u) \in S$ or $p(v) \in S$. Suppose the former holds. Since $p(x)$ is a polynomial operation of $\mathbf{F}'$, there is some term $g(x_1, x_2, \bar{y}')$ and elements $\bar{b}$ in F such that $p(x) = g^{\mathbf{F}'}(x, \mathbf{z}, \bar{b})$. Also, since $\langle u, v \rangle \in C_j$ then there is a term $t(x, \bar{y})$ which is not left invertible at x and elements $\bar{c}$ in F such that

$$\langle u, v \rangle = \langle t^{\mathbf{F}'}(0, \bar{c}), t^{\mathbf{F}'}(\mathbf{z}, \bar{c}) \rangle.$$

Since $p(u) \in S$, then we can find a term $r(x, \bar{y}'')$, not left invertible at x, and elements $\bar{d}$ in F, such that

$$\langle r^{\mathbf{F}'}(0, \bar{d}), r^{\mathbf{F}'}(\mathbf{z}, \bar{d})\rangle = \langle a, p(u)\rangle.$$

Thus we have

$$g^{\mathbf{F}'}(t^{\mathbf{F}'}(0, \bar{c}), \mathbf{z}, \bar{b}) = r^{\mathbf{F}'}(\mathbf{z}, \bar{d}).$$

Now since $\mathbf{F}'$ is free and 0, $\bar{b}$, $\bar{c}$, $\bar{d}$ are in F, then we have

$$g^{\mathbf{F}'}(t^{\mathbf{F}'}(0, \bar{c}), q, \bar{b}) = r^{\mathbf{F}'}(q, \bar{d})$$

for any $q \in F_1'$. Thus

$$g^{\mathbf{F}'}(t^{\mathbf{F}'}(0, \bar{c}), 0, \bar{b}) = r^{\mathbf{F}'}(0, \bar{d}) = a.$$

It follows from the above observations, and the strongly Abelian property, that since r is not left invertible at x, then the term

$$s(x_1, x_2, \bar{y}, \bar{y}') = g(t(x_1, \bar{y}), x_2, \bar{y}')$$

is not left invertible at x_2. Also, since $t(x, \bar{y})$ is not left invertible at x, then s is not left invertible at x_1. If the term $f(x, \bar{y}, \bar{y}') = s(x, x, \bar{y}, \bar{y}')$ were left invertible at x then we would have

$$\mathbf{A}[d] \models e(s(x, x, \bar{y}, \bar{y}'), \bar{w}) \approx x$$

for some term e of the appropriate type. Using Lemma 11.11 (i) we could then conclude that s is left invertible at either x_1 or x_2.

Therefore $\langle f^{\mathbf{F}'}(0, \bar{c}, \bar{b}), f^{\mathbf{F}'}(\mathbf{z}, \bar{c}, \bar{b})\rangle \in C_i$. But we have

$$f^{\mathbf{F}'}(0, \bar{c}, \bar{b}) = r^{\mathbf{F}'}(0, \bar{d}) = a$$

and

$$f^{\mathbf{F}'}(\mathbf{z}, \bar{c}, \bar{b}) = g^{\mathbf{F}'}(t^{\mathbf{F}'}(\mathbf{z}, \bar{c}), \mathbf{z}, \bar{b}) = p(v).$$

The proof for the remaining case, when $p(v) \in S$, is similar to the one just given. (It is left to the reader.) This concludes our demonstration that $a/\theta = S$.

To prove that $\mathbf{z}/\theta = \{\mathbf{z}\}$, we must show that if $\langle u, v\rangle \in C_j$ for some $1 \leq j \leq k$, and if p is a unary polynomial operation of $\mathbf{F}'$, then $p(u) = \mathbf{z}$ if and only if $p(v) = \mathbf{z}$. As above, we choose terms $t(x, \bar{y})$, $g(x', y, \bar{y})$ such that t is not left invertible at x and elements $\bar{b}$, $\bar{c}$ in F such that $p(x) = g^{\mathbf{F}'}(x, \mathbf{z}, \bar{b})$, and $\langle u, v\rangle = \langle t^{\mathbf{F}'}(0, \bar{c}), t^{\mathbf{F}'}(\mathbf{z}, \bar{c})\rangle$.

Now if $p(u) = \mathbf{z}$, then $g^{\mathbf{F}'}(t^{\mathbf{F}'}(0, \bar{c}), \mathbf{z}, \bar{b}) = \mathbf{z}$. Since $\mathbf{z}$ is a free generator of $\mathbf{F}'$ it follows that the term $g(t(x_1, \bar{y}), x_2, \bar{y}')$ is right invertible, and hence essentially unary. So, the equation $g(t(x_1, \bar{y}), x_2, \bar{y}') \approx x_2$ is valid in $\mathcal{V}[d]$. Thus

$$\mathbf{z} = g^{\mathbf{F}'}(t^{\mathbf{F}'}(0, \bar{c}), \mathbf{z}, \bar{b}) = g^{\mathbf{F}'}(t^{\mathbf{F}'}(\mathbf{z}, \bar{c}), \mathbf{z}, \bar{b}) = p(v).$$

On the other hand, if $p(v) = \mathbf{z}$, then $g^{\mathbf{F}'}(t^{\mathbf{F}'}(\mathbf{z}, \bar{c}), \mathbf{z}, \bar{b}) = \mathbf{z}$, and again we conclude that the term $g(t(x_1, \bar{y}), x_2, \bar{y}')$ is right invertible, and hence essentially unary. Since t is not left invertible, it follows that the equation $g(t(x_1, \bar{y}), x_2, \bar{y}') \approx x_2$ is valid in $\mathcal{V}[d]$. Thus

$$\mathbf{z} = g^{\mathbf{F}'}(t^{\mathbf{F}'}(\mathbf{z}, \bar{c}), \mathbf{z}, \bar{b}) = g^{\mathbf{F}'}(t^{\mathbf{F}'}(0, \bar{c}), \mathbf{z}, \bar{b}) = p(u),$$

which completes the proof of this lemma. □

Now we define

$$\mathbf{C}(\overline{X}, 0, \mathbf{z}) = \mathbf{F}'/\theta,$$

where $\theta = \theta(\overline{X}, 0, \mathbf{z})$. By Lemma 11.13, the image of $\mathbf{F}$ in $\mathbf{C}(\overline{X}, 0, \mathbf{z})$ under the quotient mapping is isomorphic to $\mathbf{F}$. In the following considerations, we will identify $\mathbf{F}$ with its image $\mathbf{F}/\theta$, and also identify $\mathbf{z}$ with $\mathbf{z}/\theta$ in $\mathbf{C}(\overline{X}, 0, \mathbf{z})$, since $\mathbf{z}/\theta = \{\mathbf{z}\}$. So, by construction, inside the strongly Abelian algebra $\mathbf{C} = \mathbf{C}(\overline{X}, 0, \mathbf{z})$, the following holds:

$$t^{\mathbf{C}}(0, \bar{u}) = t^{\mathbf{C}}(\mathbf{z}, \bar{u})$$

for all $\bar{u} \in C(\overline{X}, 0, \mathbf{z})$ of the appropriate sort and terms $t(x, \bar{y})$ such that t is not left invertible at x and $\sigma(x) = 1$. What this construction has accomplished is that we have properly extended the algebra $\mathbf{F}$ so that in the extension there is a new element, $\mathbf{z}$, that behaves just as 0 does with respect to certain term operations.

Notice that it follows from Lemma 11.8 (ii) that no term operation of $\mathbf{A}[d]$ depends on more than $k \cdot |A|$ variables.

Definition 11.14 Let $m = k \cdot |A|$ and let T be a finite set of terms of $\mathcal{V}[d]$, each containing at most the variables $x_1, \ldots, x_m$, such that every term operation of $\mathbf{A}[d]$ is essentially equal to one induced by a term of T. Let T_1 be the subset of T consisting of all essentially unary terms that depend only on their first variable. Let N_i be the set of terms $t(y_1, \ldots, y_s)$ in T such that $\{y_1, \ldots, y_s\} \subseteq \{x_1, \ldots, x_m\}$, $\sigma(y_1) = i$, and t is not left invertible at y_1. Let $a \sim b$ be the formula

$$(\sigma(a) = \sigma(b)) \wedge \bigwedge_{t \in \mathrm{N}_{\sigma(a)}} (\forall \bar{u})(t(a, \bar{u}) \approx t(b, \bar{u})).$$

LEMMA 11.15 *Let* $\mathbf{B}$, $\mathbf{D}_j \in \mathcal{V}[d]$ *for* $j \in I$.

(i) $\sim$ *defines an equivalence relation on* B *that is first order definable.*

(ii) *Let* $s(x, \bar{z})$ *be a right invertible term depending only on* x *with* $\sigma(x) = i$ *and let* a, $b \in B_i$ *with* $a \sim b$. *Then* $s^{\mathbf{B}}(a, \bar{c}) \sim s^{\mathbf{B}}(b, \bar{c})$ *for all* $\bar{c}$ *of the appropriate sort.*

(iii) $\mu \sim \nu$ *in* $\prod_{j \in I} \mathbf{D}_j$ *if and only if* $\mu(j) \sim \nu(j)$ *in* $\mathbf{D}_j$, *for each* $j \in I$.

PROOF. Statements (i) and (iii) are immediate from the definition.

To prove (ii), suppose that $s(x, \bar{z})$ is right invertible with $\sigma(x) = i$ and a, $b \in B_i$ with $a \sim b$. If $t(y_1, \ldots, y_s) \in N_i$ then it follows that $t^{\mathbf{B}}(s^{\mathbf{B}}(a, \bar{c}), \bar{u}) = t^{\mathbf{B}}(s^{\mathbf{B}}(b, \bar{c}), \bar{u})$ for all $\bar{c}$ and $\bar{u}$ of the appropriate sort, since the term $t(s(y_1, \bar{z}), y_2, \ldots, y_s)$ is not left invertible at y_1. This shows that $s^{\mathbf{B}}(a, \bar{c}) \sim s^{\mathbf{B}}(b, \bar{c})$. □

Proof of Theorem 11.9. Choose a term $q(x_1, \ldots, x_s)$ as in Lemma 11.12. We may assume, without loss of generality, that the sort of q is 1. Set σ_1 to $\sigma(x_1)$ and σ_2 to $\sigma(x_2)$. Let

$$\{\mathbf{x}_1, \mathbf{x}_2, \ldots, \mathbf{x}_k, \mathbf{a}, \mathbf{a}', \mathbf{b}, \mathbf{b}'\}$$

be a set of $k + 4$ distinct elements, and let $\mathbf{F}$ be the free algebra in $\mathcal{V}[d]$ generated by $\overline{X} = \langle X_i : i \leq k \rangle$, where for $i \leq k$, X_i consists of $\mathbf{x}_i$ along with $\mathbf{a}$ and $\mathbf{a}'$ if $\sigma_1 = i$, and $\mathbf{b}$ and $\mathbf{b}'$ if $\sigma_2 = i$.

Let $x \bullet y$ denote the $\mathbf{F}$ polynomial operation $q^{\mathbf{F}}(x, y, \mathbf{x}_{i_3}, \ldots, \mathbf{x}_{i_s})$, where the $\mathbf{x}_{i_j}$ are of the appropriate sort. Let $\mathbf{0} = \mathbf{a} \bullet \mathbf{b}$, $\mathbf{1} = \mathbf{a} \bullet \mathbf{b}'$, $\mathbf{2} = \mathbf{a}' \bullet \mathbf{b}$ and $\mathbf{3} = \mathbf{a}' \bullet \mathbf{b}'$. Choose a new element $\mathbf{z}$, and recalling our earlier construction, let $\mathbf{C} = \mathbf{C}(\overline{X}, \mathbf{0}, \mathbf{z})$. Then $\mathbf{C}$ is a finite member of $\mathcal{V}[d]$, which we regard as extending the algebra $\mathbf{F}$.

Since $\mathbf{a}$, $\mathbf{a}'$, $\mathbf{b}$ and $\mathbf{b}'$ are free generators of $\mathbf{F}$ and q depends on x_1 and x_2, then it follows that $\mathbf{0}$, $\mathbf{1}$, $\mathbf{2}$ and $\mathbf{3}$ are all distinct. In fact they are all distinct modulo the relation $\sim$, as the following claim shows.

Claim 1. For i, $j \in \{0, 1, 2, 3\}$ distinct, $\mathbf{C} \models (i \not\sim j)$. Also, $\mathbf{0} \sim \mathbf{z}$, $\mathbf{a} \not\sim \mathbf{a}'$ and $\mathbf{b} \not\sim \mathbf{b}'$.

$\mathbf{0} \sim \mathbf{z}$ follows by construction, using the strongly Abelian property. Since $\mathbf{a}$ and $\mathbf{a}'$ are free generators of $\mathbf{F}$ and the term q is not left invertible at x_1, then $\mathbf{a} \sim \mathbf{a}'$ in $\mathbf{C}$ would imply that q does not depend on its first variable. This is contrary to our choice of q and so we must have $\mathbf{a} \not\sim \mathbf{a}'$. Similarly, $\mathbf{b} \not\sim \mathbf{b}'$.

Now suppose that we have $\mathbf{0} \sim \mathbf{1}$ in $\mathbf{C}$. Since $\mathbf{F} \leq \mathbf{C}$ and $\mathbf{0}$, $\mathbf{1} \in F$, it follows that $\mathbf{0} \sim \mathbf{1}$ in $\mathbf{F}$ too. Let $\alpha = \mathrm{Cg}^{\mathbf{F}}(\langle \mathbf{0}, \mathbf{1} \rangle)$. We will show that

$\langle \mathbf{2}, \mathbf{3} \rangle \notin \alpha$ in order to conclude that $\mathbf{F}/\alpha$ is not Abelian, contrary to our assumptions.

In fact $\mathbf{3}/\alpha = \{\mathbf{3}\}$. If not, choose $v \in F$ such that $v \neq \mathbf{3}$, $\langle v, \mathbf{3} \rangle \in \alpha$ and $\{\mathbf{3}, v\} = p(\{\mathbf{0}, \mathbf{1}\})$ for some polynomial operation $p \in \mathrm{Pol}_1\mathbf{F}$. Let $p(x) = g^{\mathbf{F}}(x, \bar{u})$ for some $\bar{u} \in F$ and some term $g(x, \bar{y})$. Now if g is not left invertible at x, then $\mathbf{0} \sim \mathbf{1}$ implies that $g^{\mathbf{F}}(\mathbf{0}, \bar{u}) = g^{\mathbf{F}}(\mathbf{1}, \bar{u})$, i.e., $p(\mathbf{0}) = p(\mathbf{1})$, contradicting $v \neq \mathbf{3}$.

If g is left invertible at x, then the function $p(x)$ is one to one on F_1. There are two cases to consider, either $p(\mathbf{0}) = \mathbf{3}$ or $p(\mathbf{1}) = \mathbf{3}$.

If $p(\mathbf{0}) = \mathbf{3}$, then we have $g^{\mathbf{F}}(\mathbf{a} \bullet \mathbf{b}, \bar{u}) = \mathbf{a}' \bullet \mathbf{b}'$. Since $\mathbf{F}$ is free and strongly Abelian and the free generator $\mathbf{a}$ does not occur on the right side of this equality, it follows that $g^{\mathbf{F}}(\mathbf{a}' \bullet \mathbf{b}, \bar{u}) = \mathbf{a}' \bullet \mathbf{b}'$, i.e., $p(\mathbf{2}) = \mathbf{3}$. This contradicts that p is one to one. A similar argument shows that $p(\mathbf{1}) = \mathbf{3}$ is also impossible. Thus $\mathbf{3}/\alpha = \{\mathbf{3}\}$ and, in particular, $\langle \mathbf{2}, \mathbf{3} \rangle \notin \alpha$.

But then

$$\mathbf{F}/\alpha \models (\mathbf{a}/\alpha) \bullet (\mathbf{b}/\alpha) = (\mathbf{a}/\alpha) \bullet (\mathbf{b}'/\alpha)$$

and

$$\mathbf{F}/\alpha \models (\mathbf{a}'/\alpha) \bullet (\mathbf{b}/\alpha) \neq (\mathbf{a}'/\alpha) \bullet (\mathbf{b}'/\alpha),$$

and so $\mathbf{F}/\alpha$ is not Abelian, which is impossible. The remaining cases are handled similarly.

Recall from Chapter 0 that a graph $\mathbf{G} = \langle G, E \rangle$ is called bi-partite if and only if $G = G_1 \cup G_2$ for some disjoint nonempty subsets G_1 and G_2 with $E \subseteq G_1 \times G_2 \cup G_2 \times G_1$. Let $\mathcal{B}$ be the class of bi-partite graphs. We leave it as an exercise to show that $\mathcal{B}$ is ω-unstructured.

We will interpret $\mathcal{B}$ into $\mathcal{V}[d]$ in order to prove our theorem. We use the first order formalization of $\mathcal{V}[d]$ mentioned earlier to construct our interpretation. So, we have at our disposal k unary predicates $\{\mathrm{U}_i(x) : 1 \leq i \leq k\}$ such that for any $\mathbf{B} \in \mathcal{V}[d]$, if $b \in B$, then

$$\sigma(b) = i \quad \text{if and only if} \quad \mathbf{B} \models \mathrm{U}_i(b).$$

Let $\mathbf{G} = \langle G, E \rangle \in \mathcal{B}$, say $E \subseteq G_1 \times G_2 \cup G_2 \times G_1$ for certain disjoint nonempty subsets G_1 and G_2 of G with $G = G_1 \cup G_2$. Let

$$\widehat{E} = \{\{u, v\} : \langle u, v \rangle \in E\}.$$

Choose two new points p_1 and p_2 and let $Y = G \cup \{p_1, p_2\}$.

For $v \in G_1$, let $f_v : Y \to C$ be defined by

$$f_v(x) = \begin{cases} \mathbf{a}' & \text{if } x = v \\ \mathbf{a} & \text{otherwise.} \end{cases}$$

For $v \in G_2$, let $f_v : Y \to C$ be defined by

$$f_v(x) = \begin{cases} \mathbf{b}' & \text{if } x = v \\ \mathbf{b} & \text{otherwise.} \end{cases}$$

For $i \in \{1,2\}$ and $e = \{v, w\} \in \widehat{E}$, where $v \in G_1$, $w \in G_2$, let $f_e^i : Y \to C$ be defined by

$$f_e^i(x) = \begin{cases} \mathbf{2} & \text{if } x = v \\ \mathbf{1} & \text{if } x = w \\ \mathbf{z} & \text{if } x = p_i \\ \mathbf{0} & \text{otherwise.} \end{cases}$$

Let

$$\begin{aligned} G_1^* &= \{f_v : v \in G_1\}, \\ G_2^* &= \{f_v : v \in G_2\}, \end{aligned}$$

and

$$E^* = \{f_e^i : e \in \widehat{E} \text{ and } i \in \{1,2\}\}.$$

We define $\mathbf{B} \subseteq \mathbf{C}^Y$ as follows:

$\mathbf{B}$ is the subalgebra of $\mathbf{C}^Y$
generated by the set $G_1^* \cup G_2^* \cup E^* \cup \{\hat{\mathbf{x}}_j : j \geq 1\}$,

where for $d \in C$, $\hat{d}$ denotes the constant valued function in C^Y with value d. Throughout this proof, the elements of $G_1^* \cup G_2^* \cup E^* \cup \{\hat{\mathbf{x}}_j : j \geq 1\}$ will be referred to as generators.

We will now attempt to recover (as best we can) the sets G_1^*, G_2^* and E^* in $\mathbf{B}$ using first order formulas (with parameters).

Claim 2. Let $v \in G$, $e \in \widehat{E}$ and $i \in \{1,2\}$. If f_v or f_e^i equals $t^{\mathbf{B}}(\mu_1, \ldots, \mu_s)$ for some term t and elements $\mu_1, \ldots, \mu_s \in B$, then the term t is right invertible and hence is essentially unary.

To prove this, first suppose that $f_v = t^{\mathbf{B}}(\mu_1, \ldots, \mu_s)$ as above, and assume that $v \in G_1$. (The proof for the case $v \in G_2$ can easily be reconstructed from the argument we shall now present.) Then, at the coordinate $v \in Y$, we have $\mathbf{a}' = t^{\mathbf{C}}(\mu_1(v), \ldots, \mu_s(v))$. Since at the coordinate v all of the generators of $\mathbf{B}$ are in F, then for all $1 \leq i \leq s$, $\mu_i(v)$ is in F too. Thus we have $\mathbf{a}' = t^{\mathbf{F}}(\mu_1(v), \ldots, \mu_s(v))$ holding in the free algebra $\mathbf{F}$. But then it follows that t is right invertible since $\mathbf{a}'$ is a free generator of $\mathbf{F}$. So, by Lemma 11.11, it is essentially unary.

If $f_e^i = t^{\mathbf{B}}(\mu_1, \ldots, \mu_s)$ then at $p_i \in Y$ we have $\mathbf{z} = t^{\mathbf{C}}(\mu_1(p_i), \ldots, \mu_s(p_i))$ and so

$$\langle \mathbf{z}, t^{\mathbf{F}'}(u_1, \ldots, u_s)\rangle \in \theta(\overline{X}, \mathbf{0}, \mathbf{z}) = \theta$$

for some $u_j \in F'$ with $u_j/\theta = \mu_j(p_i)$ for $j \leq s$. But then by Lemma 11.13,

$$\mathbf{z} = t^{\mathbf{F}'}(u_1, \ldots, u_s)$$

and hence t is right invertible since $\mathbf{z}$ is a free generator of $\mathbf{F}'$.

Since the algebra $\mathbf{C}$ is finite then there is a first order formula with parameters, $\mathrm{Con}(x)$, such that

$$\mathbf{B} \models \mathrm{Con}(\mu) \text{ if and only if } \mu \text{ is constant valued.}$$

Also, since the set of terms T from Definition 11.14 is finite then the following property is expressible by a first order formula, $\mathrm{Gen}(x)$:

> $\neg\mathrm{Con}(x)$ and x is not in the range of any term in T that is not right invertible.

From Claim 2 we have that

$$\mathbf{B} \models \mathrm{Gen}(f)$$

for any nonconstant generator f.

Another first order property that the nonconstant generators satisfy can be described in terms of the following quasi-order and related equivalence relation on B:

> for a and b in B we write $a \leq b$ if there exists an essentially unary term $t(x, \bar{y})$ in T_1 such that $a = t^{\mathbf{B}}(b, \bar{c})$ for some $\bar{c}$, and write $a \equiv b$ if $a \leq b$ and $b \leq a$.

Since T_1 is finite these two relations are first order definable.

We let $\mathrm{Max}(x)$ be the formula

$$\mathrm{Gen}(x) \wedge \forall y\, [(\mathrm{Gen}(y) \wedge x \leq y) \rightarrow y \leq x].$$

Claim 3. $\mathbf{B} \models \mathrm{Max}(f)$ for all nonconstant generators f. If $\mathbf{B} \models \mathrm{Max}(\lambda)$ then $\lambda \equiv f$ for some unique nonconstant generator f.

Both parts of this claim rest on the fact that if f and g are nonconstant generators and $f \leq g$ then $f = g$. There are several cases to consider in proving this. First of all, if $f \leq g$, then $f = t^{\mathbf{B}}(g)$ for some essentially unary term $t(x)$ in T_1, where for convenience we have listed only the variable x that t depends on. Now if $g = f_e^i$ for some i and e and $f = f_u$ for some $u \in G$, then at the coordinate p_j, where $i \neq j$, we get that $t^{\mathbf{F}}(f_e^i(p_j))$

is equal to either $\mathbf{a}$ or $\mathbf{b}$, depending on whether u is in G_1 or G_2. Since $f_e^i(p_j) = \mathbf{0} = \mathbf{a} \bullet \mathbf{b}$ then this implies that the term q is left invertible in one of the variables x_1 or x_2. This is a contradiction.

If f equals $f_{e'}^j$ for some j and e' and j is not equal to i, then evaluating $f = t^{\mathbf{B}}(g)$ at the coordinate p_j yields $t^{\mathbf{C}}(\mathbf{0}) = \mathbf{z}$. This is impossible since F is a proper subuniverse of $\mathbf{C}$ that contains $\mathbf{0}$ and not $\mathbf{z}$. Thus $j = i$.

Suppose that $e \neq e'$, say $v \in e - e'$. Without loss of generality, assume that v is in G_1. Then evaluating $t^{\mathbf{B}}(g) = f$ at the coordinate v, we have $t^{\mathbf{F}}(\mathbf{a}' \bullet \mathbf{b}) = \mathbf{a} \bullet \mathbf{b}$. This implies that the term q is independent of its first variable, since $\mathbf{a}$, $\mathbf{a}'$ and $\mathbf{b}$ are free generators of $\mathbf{F}$. This is contrary to our choice of q.

The situation when the generator g is equal to f_u for some u in G can be dispensed with in a similar fashion, and is left to the reader.

To show that $\mathbf{B} \models \mathrm{Max}(f)$ for any nonconstant generator f, we must show that if $f \leq \mu$ for some $\mu \in B$ with $\mathbf{B} \models \mathrm{Gen}(\mu)$, then $\mu \leq f$. If $\mathbf{B} \models \mathrm{Gen}(\mu)$ then μ is in B and so is equal to some term operation of $\mathbf{B}$ applied to some generators. This term operation must be essentially unary since $\mathbf{B} \models \mathrm{Gen}(\mu)$ and from this it easily follows that $\mu \leq g$ for some nonconstant generator g. Since $f \leq \mu$, then by transitivity, $f \leq g$ and so $f = g$. Thus, $\mu \leq f$.

The above argument also shows that if $\mathbf{B} \models \mathrm{Max}(\lambda)$, then $\lambda \equiv f$ for some unique nonconstant generator f.

The following first order formulas are used to distinguish between the f_v's and f_e^i's up to the definable equivalence relation $\equiv$. Let $\mathrm{Edg}(x)$ be the formula

$$\mathrm{Max}(x) \wedge \exists x', y\,(\mathrm{Max}(x') \wedge \mathrm{Max}(y) \wedge x \equiv x' \wedge x \not\equiv y \wedge x' \sim y)$$

and let $\mathrm{Un}(x)$ be

$$\mathrm{Max}(x) \wedge \neg\mathrm{Edg}(x).$$

Claim 4. For $\mu \in B$ with $\mathbf{B} \models \mathrm{Max}(\mu)$,

$$\mathbf{B} \models \mathrm{Edg}(\mu) \quad \text{if and only if} \quad \mu \equiv f_e^i$$

for some $i \in \{1, 2\}$ and $e \in \widehat{E}$ and

$$\mathbf{B} \models \mathrm{Un}(\mu) \quad \text{if and only if} \quad \mu \equiv f_u$$

for some u in G.

By construction, $\mathbf{B} \models \mathrm{Edg}(f_e^i)$ since $f_e^1 \not\equiv f_e^2$ and $f_e^1 \sim f_e^2$.

To finish this claim we must show that $\mathbf{B} \not\models \mathrm{Edg}(f_u)$ for any u in G. If not, then we would have some μ and ν in B with

$$\mathbf{B} \models \mathrm{Max}(\mu) \wedge \mathrm{Max}(\nu) \wedge f_u \equiv \mu \wedge f_u \not\equiv \nu \wedge \mu \sim \nu.$$

Since $\mathrm{Max}(\nu)$, then $\nu \equiv g$ for some nonconstant generator g, and $f_u \not\equiv \nu$ implies that $f \neq g$.

The above equivalences imply that there must be right invertible (essentially) unary terms $r(x)$, $s(x)$ and $t(x)$ such that $r^{\mathbf{B}}(\nu) = g$, $s^{\mathbf{B}}(g) = \nu$ and $t^{\mathbf{B}}(f_u) = \mu$.

Applying Lemma 11.15 to the equivalence $\mu \sim \nu$ and the term $r(x)$, we conclude that $r^{\mathbf{B}}(\mu) \sim r^{\mathbf{B}}(\nu)$. Thus $(rt)^{\mathbf{B}}(f_u) \sim g$.

If $g = f_v$ for some $v \neq u$, then by Lemma 11.15 (iii), $f_v(v) \sim f_v(p_1)$, since $f_u(v) = f_u(p_1)$. This implies that either $\mathbf{a}' \sim \mathbf{a}$ or $\mathbf{b}' \sim \mathbf{b}$ depending on whether v is in G_1 or G_2. In either case we contradict Claim 1.

If $g = f_e^i$ for some $e \in \widehat{E}$, then choosing v in $e - \{u\}$, say $v \in G_1$, and using Lemma 11.15 once more, we conclude that $f_e^i(v) \sim f_e^i(p_i)$ since $f_u(v) = f_u(p_i)$. This gives $\mathbf{2} \sim \mathbf{z} \sim \mathbf{0}$, which contradicts Claim 1.

The last step in the proof is to recover the binary graph relation. Recall that the sorts of the variables x_1 and x_2 of the term q are σ_1 and σ_2 respectively. For convenience we will also use the symbol $\bullet$ to denote the the binary operation $q^{\mathbf{B}}(x, y, \hat{\mathbf{x}}_{i_3}, \ldots, \hat{\mathbf{x}}_{i_s})$. Let $\mathrm{E}(x, y)$ be the formula

$$\begin{gathered}\mathrm{Un}(x) \wedge \mathrm{Un}(y) \wedge \exists x', y', w \, (\mathrm{Un}(x') \wedge \mathrm{Un}(y') \wedge \\ \mathrm{Edg}(w) \wedge x \equiv x' \wedge y \equiv y' \wedge \\ (w \sim x' \bullet y' \vee w \sim y' \bullet x')).\end{gathered}$$

Claim 5. For μ, $\nu \in B$ such that $\mathbf{B} \models \mathrm{Un}(\mu) \wedge \mathrm{Un}(\nu)$,

$$\mathbf{B} \models \mathrm{E}(\mu, \nu) \text{ if and only if } \mu \equiv f_u \text{ and } \nu \equiv f_v$$

for some u, $v \in G$ with $\langle u, v \rangle \in E$.

The direction from right to left is easy to establish, since $\mathbf{B} \models f_e^1 \sim f_u \bullet f_v$ whenever $e = \{u, v\} \in \widehat{E}$ and $u \in G_1$, $v \in G_2$.

For the converse, suppose that $\mathbf{B} \models \mathrm{E}(\mu, \nu)$, say the elements μ', ν' and ϵ witness this. Then for some u, v in G we have $\mu \equiv \mu' \equiv f_u$, $\nu \equiv \nu' \equiv f_v$, $\mathbf{B} \models \mathrm{Edg}(\epsilon)$ and $\epsilon \sim \mu' \bullet \nu'$ or $\epsilon \sim \nu' \bullet \mu'$. Without loss of generality $\epsilon \sim \mu' \bullet \nu'$.

Since $\mathbf{B} \models \mathrm{Edg}(\epsilon)$, there is some $e \in \widehat{E}$ and $i \leq 2$ such that $\epsilon \equiv f_e^i$. So there are right invertible (essentially) unary terms $q(x)$, $r(x)$, $s(x)$ and

$t(x)$ with $q^{\mathbf{B}}(\epsilon) = f_e^i$, $r^{\mathbf{B}}(f_e^i) = \epsilon$, $s^{\mathbf{B}}(f_u) = \mu'$ and $t^{\mathbf{B}}(f_v) = \nu'$. By Lemma 11.15 the relation $\epsilon \sim \mu' \bullet \nu'$ implies that $q^{\mathbf{B}}(\epsilon) \sim q^{\mathbf{B}}(\mu' \bullet \nu')$, i.e., $f_e^i \sim q^{\mathbf{B}}(s^{\mathbf{B}}(f_u) \bullet t^{\mathbf{B}}(f_v))$.

Now, if $e \neq \{u, v\}$, then choose $x \in e$ different from u and v. Since $f_u(x) = f_u(p_i)$ and $f_v(x) = f_v(p_i)$ it follows, using Lemma 11.15, that $f_e^i(x) \sim f_e^i(p_i)$. This implies that $\mathbf{2} \sim \mathbf{z} \sim \mathbf{0}$ or $\mathbf{1} \sim \mathbf{z} \sim \mathbf{0}$, depending on whether x is in G_1 or G_2. This contradicts Claim 1.

Thus we have established that the structure

$$\langle \mathrm{Un}^{\mathbf{B}}, \mathrm{E}^{\mathbf{B}} \rangle / {\equiv^{\mathbf{B}}}$$

is isomorphic to our original graph $\mathbf{G} = \langle G, E \rangle$. It is important to observe that if $\mathbf{G}$ is finite then $\mathbf{B}$ is finite. This point is what allows us to conclude that $\mathcal{V}$ is ω-unstructured. □

COROLLARY 11.16 *Let $\mathbf{A} = \langle A, f \rangle$ be a finite strongly Abelian algebra with one fundamental operation. Then $\mathbf{V}(\mathbf{A})$ is decidable (structured) if and only if f is δ-injective or essentially unary.*

PROOF. One can show by an easy induction that if f is not δ-injective, then the only δ-injective operations in Clo $\mathbf{A}$ are the projection operations. Thus, if $\mathbf{V}(\mathbf{A})$ is decidable, and if f is not δ-injective, then f must be essentially unary.

Conversely, if f is essentially unary then $\mathbf{V}(\mathbf{A})$ is decidable, since any variety of mono-unary algebras is decidable. If f is δ-injective, then using Theorem 4.9 of R. McKenzie [1983], we again conclude that $\mathbf{V}(\mathbf{A})$ is decidable. □

We will see in the next chapter that in order for $\mathcal{V}$ to be decidable, the term operations of $\mathbf{A}$ must also satisfy a very special condition referred to as linearity.

Chapter 12

The unary case

In this chapter we will find necessary and sufficient conditions for a locally finite, strongly Abelian variety $\mathcal{V}$ to be decidable. It will turn out that for such varieties, the properties of being undecidable, hereditarily undecidable, unstructured, or ω-unstructured, all coincide. The results of Chapter 11 reduce the problem to determining those locally finite, essentially unary, k-sorted varieties (for $k \geq 1$) which are decidable. For the purposes of determining the decidability of $\mathcal{V}$, we may assume that $\mathcal{V}$ is in fact a multi-unary k-sorted variety of finite type. Thus each term in the language of $\mathcal{V}$ has at most one variable.

Definition 12.1 Let $\mathcal{W}$ be a multi-unary k-sorted variety (for $k \geq 1$) of finite type.

(1) A term $t(x)$ of $\mathcal{W}$ is said to be **constant** if

$$\mathcal{W} \models t(x) \approx t(y),$$

where x and y are distinct variables of the appropriate sort.

(2) We say that $\mathcal{W}$ is **linear** if for all nonconstant terms $t(x)$, $s(y)$ of $\mathcal{W}$ with the variables x and y of the same sort, there is a term $w(z)$ of $\mathcal{W}$ of the appropriate type such that either the equation

$$t(x) \approx w(s(x))$$

or the equation

$$s(x) \approx w(t(x))$$

holds in $\mathcal{W}$.

THEOREM 12.2 *Let $k \geq 1$ and let $\mathcal{V}$ be a locally finite, multi-unary, k-sorted variety of finite type. Then $\mathcal{V}$ is decidable (structured) if and only if it is linear.*

In the proof of Theorem 12.2 contained in this chapter, we assume throughout that $\mathcal{V}$ is an ordinary (1-sorted) variety, i.e., that $k = 1$. The proof for $k > 1$ is essentially the same, involving no new tricks, but it is a little messier to present. The reader may also consult M. Valeriote [1987] for details of the proof which we will present in this chapter. In the 1-sorted case we can associate a finite monoid to $\mathcal{V}$ the structure of which determines the decidability of $\mathcal{V}$.

Definition 12.3 Let $\mathcal{V}$ be a multi-unary (1-sorted) variety of finite type, and let $\mathbf{F} = \mathbf{F}_{\mathcal{V}}(\mathbf{x})$ be the $\mathcal{V}$-free algebra on one generator. Let $\mathbf{M}(\mathcal{V}) = \langle F, \circ, 1\rangle$ be the monoid with universe F and multiplication $\circ$ defined by:

$$m(\mathbf{x}) \circ n(\mathbf{x}) = m(n(\mathbf{x})).$$

Let $1 = \mathbf{x}$.

We henceforth assume that $\mathcal{V}$ is a locally finite, multi-unary, 1-sorted variety of finite type. Notice that $\mathbf{M}(\mathcal{V})$ is finite since $\mathcal{V}$ is locally finite. For convenience, we will not distinguish between elements of $\mathbf{M}(\mathcal{V})$ and the corresponding terms in the language. Often, elements of $\mathbf{M}(\mathcal{V})$ will be regarded as actual terms, especially in the first order formulas constructed for our interpretations.

The linearity of our 1-sorted variety $\mathcal{V}$ can be determined by examining the ordering of right-sided divisibility on the finite monoid $\mathbf{M} = \mathbf{M}(\mathcal{V})$. Let $G \subseteq M$ be the set of invertible elements of $\mathbf{M}$. It is clear that G is nonempty and that $\mathbf{G}$ is a submonoid of $\mathbf{M}$.

Definition 12.4 **(1)** For $x, y \in M$, we write $x \leq y$ if for some $w \in M$, $x = wy$, and we write $x \propto y$ if for some $g \in G$, $x \leq yg$.

(2) For $x, y \in M$, we write $x \equiv y$ if $x \leq y$ and $y \leq x$, and $x \sim y$ if $x \propto y$ and $y \propto x$.

(3) $\mathbf{M}$ is said to be **linear (quasi-linear)** if for every $x, y \in M$, $x \leq y$ or $y \leq x$ ($x \propto y$ or $y \propto x$).

LEMMA 12.5

(i) M *is linear if and only if* $\mathcal{V}$ *is linear.*

(ii) *Both* $\leq$ *and* $\propto$ *define transitive, reflexive relations on* M.

(iii) *Both* $\equiv$ *and* $\sim$ *define equivalence relations on* M. *Modulo the respective equivalence relations,* $\leq$ *and* $\propto$ *define partial orderings.*

(iv) M *is linear (quasi-linear) if and only if* $\leq$ *modulo* $\equiv$ ($\propto$ *modulo* $\sim$) *is a linear ordering.*

(v) *For* x, $y \in M$, *if* $x \leq y$, *then* $x \propto y$.

(vi) *If* $m \in M$ *is left or right invertible, then* m *is invertible.*

PROOF. The proof of this lemma is straightforward, and so will not be given. □

Definition 12.6 For $x \in M$, let $\mathbf{G}_x = \{g \in G : x \leq xg\}$.

LEMMA 12.7

(i) *For each* $x \in M$, $\mathbf{G}_x$ *is a subgroup of* **G**; *and if* $g \in G_x$ *then* $x \equiv xg \equiv xg^{-1}$.

(ii) M *is a linear monoid if and only if* M *is quasi-linear and* $G_x = G$ *for all* $x \in M$.

PROOF.

(i) This follows from the fact that G is finite. Any nonvoid subset of a finite group, closed under multiplication, is a subgroup.

(ii) Suppose that M is linear. Then M is quasi-linear, since $\propto$ extends $\leq$. If $x \in M$, and $g \in G$, we have either $x \leq xg$ or $xg \leq x$. If the former holds then $g \in G_x$, and if the latter holds then $xg = wx$ for some $w \in M$, and so $x \leq xg^{-1}$. Thus $g^{-1} \in G_x$, and as G_x is a subgroup of G, we must also have that $g \in G_x$. Therefore $G_x = G$.

Suppose that M is quasi-linear and $G_x = G$ for all $x \in M$. Let x, $y \in M$ and assume, without loss of generality, that $x \propto y$. Then for some $g \in G$, $x \leq yg$, and as $yg \leq y$, we conclude that $x \leq y$. Thus M is linear. □

We will now give the first half of the proof of Theorem 12.2. That is, we shall show that if $\mathcal{V}$ is linear, then $\mathcal{V}$ is decidable. Until otherwise stated, we shall assume that $\mathbf{M}$ $(= \mathbf{M}(\mathcal{V}))$ is linear. Let $|M| = \kappa$.

Definition 12.8 Let $\mathbf{A} \in \mathcal{V}$ and let $a \in A$. The **orbit of** a **in** $\mathbf{A}$, is defined by

$$O_a = \{m^{\mathbf{A}}(a) : m \in M\}.$$

$\mathbf{O}_a$ will denote the subalgebra of $\mathbf{A}$ with universe O_a.

LEMMA 12.9 *Suppose that* $\mathbf{A} \in \mathcal{V}$ *and* $a \in A$.

(i) *If* $B \subseteq O_a$ *is a nonempty subuniverse of* $\mathbf{A}$, *then* $B = O_b$ *for some* $b \in B$.

(ii) *The set of subuniverses of* $\mathbf{A}$ *contained in* O_a *is linearly ordered by inclusion.*

PROOF.

(i) If $B \subseteq O_a$ is a nonempty subuniverse, then for some left ideal N of M,

$$B = \{n^{\mathbf{A}}(a) : n \in N\}.$$

Choose $m \in N$ maximal with respect to $\leq$. Then $B = O_{m(a)}$, for if $c \in B$ then $c = n(a)$ for some $n \leq m$ in N, and so $n = wm$ for some $w \in M$. Then

$$c = n(a) = (wm)(a) = w(m(a)) \in O_{m(a)}.$$

Clearly $O_{m(a)} \subseteq B$, and so $B = O_{m(a)}$.

(ii) Now let B, C be nonempty subuniverses of $\mathbf{A}$ contained in O_a, say $B = O_b$ and $C = O_c$ for some $b \in B$ and $c \in C$. Since b, $c \in O_a$, there are m, $n \in M$ with $b = m(a)$ and $c = n(a)$. If $m \leq n$, then it follows that $b \in O_c$, and thus $B \subseteq C$. $n \leq m$ leads to $C \subseteq B$. □

Definition 12.10 Let B be a nonempty set, $\mathbf{A} \in \mathcal{V}$, and $a \in A$.

(1) Let $f : B \to B$ be a function on B. For b, $c \in B$, we write $b \xrightarrow{f} c$ if $c = f^n(b)$ for some $n \geq 0$. We say $b \in B$ is **f-initial** if for all $c \in B$, $c \xrightarrow{f} b$ implies $c = b$.

(2) A function $f : O_a \to O_a$ is a **coding function for $\mathbf{O}_a$** if

(a) for all $b \in O_a$, $f(O_b) \subseteq O_b$, and

(b) the relation $\xrightarrow{f}$ on O_a is a linear ordering.

(3) A function $f : A \to A$ is a **coding function for $\mathbf{A}$** if for all $a \in A$, $f(O_a) \subseteq O_a$, and f restricted to O_a is a coding function for $\mathbf{O}_a$.

To establish the decidability of $\mathcal{V}$, we will interpret $\mathcal{V}$ into the variety of all mono-unary algebras, and so each algebra of $\mathcal{V}$ will be encoded by a single unary function on some set. The coding functions just defined will be used to construct our interpretation.

We present a few elementary facts about coding functions, without proof, in the next lemma.

LEMMA 12.11 *Let $\mathbf{A} \in \mathcal{V}$ be one-generated.*

(i) *There is a coding function for $\mathbf{A}$.*

(ii) *If $b \in A$, and $g : O_b \to O_b$ is a coding function for $\mathbf{O}_b$, then g can be extended to a coding function for $\mathbf{A}$. Conversely, if $f : A \to A$ is a coding function for $\mathbf{A}$, then f restricted to O_b is a coding function for $\mathbf{O}_b$.*

(iii) *If f is a coding function for $\mathbf{A}$, and $a' \in A$ is the unique f-initial element of A, then $O_{a'} = A$.*

(iv) *If f is a coding function for $\mathbf{A}$, and $b, c \in A$ with $b \xrightarrow{f} c$, then $c \in O_b$.*

LEMMA 12.12 *Every algebra in $\mathcal{V}$ has a coding function.*

PROOF. Let $\mathbf{A} \in \mathcal{V}$. Our proof will involve the use of Zorn's Lemma, since the algebra $\mathbf{A}$ may be infinite. Let

$$K = \{(\mathbf{B}, g) : \mathbf{B} \subseteq \mathbf{A} \text{ and } g \text{ is a coding function for } \mathbf{B}\}.$$

For $(\mathbf{B}_1, g_1), (\mathbf{B}_2, g_2) \in K$ we put

$$(\mathbf{B}_1, g_1) \subset (\mathbf{B}_2, g_2) \text{ iff } B_1 \subset B_2 \text{ and } g_2|_{B_1} = g_1.$$

This defines a partial ordering on K such that any ascending chain has an upper bound in K, namely the union of the chain. Thus by Zorn's Lemma, there is $(\mathbf{B}, f)$ in K maximal with respect to $\subset$. The algebra $\mathbf{B}$ must be equal to $\mathbf{A}$, for if $a \in A - B$, we can extend g to a coding function on the subalgebra $\mathbf{B}'$ with universe $B \cup O_a$ as follows.

If $B \cap O_a = \emptyset$, then for any coding function h for for $\mathbf{O}_a$, if we set $f' = f \cup h$ then we see that f' is a coding function for $\mathbf{B}'$, and so $(\mathbf{B}, f) \subset (\mathbf{B}', f')$ in K, contrary to our choice of $(\mathbf{B}, f)$.

If $B \cap O_a \neq \emptyset$, then $B \cap O_a = O_c$ for some $c \in B$, by Lemma 12.9. Since $f|_{O_c}$ is a coding function for $\mathbf{O}_c$, we can extend it to a coding function h for $\mathbf{O}_a$. Setting $f' = f \cup h$, we again conclude that f' is a coding function for $\mathbf{B}'$, and that $(\mathbf{B}, f) \subset (\mathbf{B}', f')$ in K.

Therefore $\mathbf{B} = \mathbf{A}$, and $f : A \to A$ is a coding function for $\mathbf{A}$. □

LEMMA 12.13 *Let $\mathbf{A} \in \mathcal{V}$ be one-generated, and let $h : A \to A$ be a coding function for $\mathbf{A}$, with h-initial element a. For each $m \in M$, there is a first order formula $\Phi^h_m(x; y_1, y_2)$ in the language of one unary operation such that for all a_1, $a_2 \in A$:*

$$\mathbf{A} \models m(a_1) = a_2 \quad \textit{if and only if} \quad \langle A, h\rangle \models \Phi^h_m(a; a_1, a_2).$$

Furthermore, if $\mathbf{B} \in \mathcal{V}$ is one-generated, $g : B \to B$ is a coding function for $\mathbf{B}$, with g-initial element b, and the structures $\langle \mathbf{A}, h\rangle$ and $\langle \mathbf{B}, g\rangle$ are isomorphic, then for any b_1, $b_2 \in B$:

$$\mathbf{B} \models m(b_1) = b_2 \quad \textit{if and only if} \quad \langle B, g\rangle \models \Phi^h_m(b; b_1, b_2).$$

PROOF. We define some auxiliary formulas. Let

$$\mathrm{d}_0(x, y) \quad \overset{\text{def}}{\longleftrightarrow} \quad x \approx y$$

and for $i \geq 0$, let

$$\mathrm{d}_{i+1}(x, y) \quad \overset{\text{def}}{\longleftrightarrow} \quad \neg \mathrm{d}_i(x, y) \wedge \mathrm{d}_i(f(x), y).$$

Thus for u, $v \in A$, $(A, h) \models \mathrm{d}_i(u, v)$ for some i if and only if $u \xrightarrow{h} v$. Also, for each $u \in A$, there is a unique number, call it $\sigma(u)$, such that

$$\langle A, h\rangle \models \mathrm{d}_{\sigma(u)}(a, u).$$

We can now define $\Phi^h_m(x; y_1, y_2)$ to be

$$\bigvee_{u \in A} \left(\mathrm{d}_{\sigma(u)}(x, y_1) \wedge \mathrm{d}_{\sigma(m(u))}(x, y_2)\right).$$

This is indeed a first order formula in the language of one unary operation, since A is finite, and $\{\sigma(u) : u \in A\}$ is a finite collection of natural numbers.

This formula does the job since for u, $v \in A$,

$$\langle A, h\rangle \models \mathrm{d}_{\sigma(u)}(a, v) \text{ if and only if } u = v,$$

and so,

$$\langle A, h\rangle \models \Phi^h_m(a; u, v) \text{ if and only if } m(u) = v.$$

If the structure $\langle \mathbf{B}, g\rangle$ is isomorphic to $\langle \mathbf{A}, h\rangle$ via the map ϕ, and b is the g-initial element, then $\phi(b) = a$, and for any $m \in M$ and b_1, $b_2 \in B$, we have:

$$\begin{aligned} \mathbf{B} \models m(b_1) = b_2 \quad & \text{if and only if} \quad \mathbf{A} \models m(\phi(b_1)) = \phi(b_2) \\ & \text{if and only if} \quad \langle A, h\rangle \models \Phi^h_m(a; \phi(b_1), \phi(b_2)) \\ & \text{if and only if} \quad \langle B, g\rangle \models \Phi^h_m(b; b_1, b_2). \end{aligned}$$

□

Since $\mathcal{V}$ is a locally finite variety, the $\mathcal{V}$-free algebra on one generator is finite, and in fact is equal in size to M. So, up to isomorphism, there are only finitely many one-generated algebras in $\mathcal{V}$. It follows that up to isomorphism there are only finitely many structures of the type $\langle \mathbf{A}, f\rangle$, where $\mathbf{A} \in \mathcal{V}$ is one-generated, and $f : A \to A$ is a coding function for $\mathbf{A}$. Let

$$\langle \mathbf{A}_1, f_1\rangle, \ldots, \langle \mathbf{A}_l, f_l\rangle$$

be a collection of these structures so that if $\langle \mathbf{B}, g\rangle$ is also such a structure, then it is isomorphic to $\langle \mathbf{A}_i, f_i\rangle$ for some unique $i \leq l$. We say that the **type of g as a coding function** is i.

LEMMA 12.14 *$\mathcal{V}$ is a decidable variety.*

PROOF. We'll construct formulas that give an interpretation of $\mathcal{V}$ into the decidable variety of mono-unary algebras. Since $\mathcal{V}$ is finitely axiomatizable (by Theorem 0.17), it will follow by Theorem 0.2 that $\mathcal{V}$ is decidable. Let $\mathbf{A} \in \mathcal{V}$ and let $f : A \to A$ be a coding function for $\mathbf{A}$. To make the presentation neater, assume that $A \cap (A \times \omega) = \emptyset$.

Setting I to be the set of f-initial elements of A, we see that I is a generating set for $\mathbf{A}$, since for all $b \in A$, $a \xrightarrow{f} b$ for some $a \in I$, implying $b \in O_a$. In fact I is a minimal generating set for $\mathbf{A}$.

For each $a \in I$, $f|_{O_a}$ is a coding function for O_a, and so has some type $i \leq l$, which we'll denote by $\mathrm{typ}(a)$. Let

$$\mathrm{Fl}(a) = \{\langle a, j\rangle : 1 \leq j \leq \mathrm{typ}(a)\},$$

and let

$$A' = A \cup \bigcup_{a \in I} \mathrm{Fl}(a).$$

Extend f to a function $f' : A' \to A'$ by

$$f'(\langle a, j\rangle) = a \ , \quad \text{for} \quad a \in I \text{ and } j \leq \mathrm{typ}(a).$$

$\mathbf{A}' = \langle A', f'\rangle$ is the mono-unary algebra into which we will interpret $\mathbf{A}$.

Consider the following formulas:

$$\begin{aligned}
\mathrm{Un}(x) &\overset{\text{def}}{\longleftrightarrow} \exists y\, (f(y) \approx x)\,; \\
\mathrm{Flag}(x) &\overset{\text{def}}{\longleftrightarrow} \neg \mathrm{Un}(x); \\
\mathrm{Init}(x) &\overset{\text{def}}{\longleftrightarrow} \exists y\, (\mathrm{Flag}(y) \wedge f(y) \approx x)\,;
\end{aligned}$$

for $1 \leq i \leq l$,

$$\begin{aligned}
\mathrm{Tp}_i(x) &\overset{\text{def}}{\longleftrightarrow} \mathrm{Init}(x) \wedge \exists y_1, \ldots, y_i \Big[\big(\bigwedge_{1 \leq j \leq i} f(y_j) \approx x\big) \\
&\qquad \wedge \big(\bigwedge_{1 \leq j < k \leq i} y_j \not\approx y_k\big) \wedge \forall z \big(f(z) \approx x \to \bigvee_{1 \leq j \leq i} z \approx y_j\big)\Big]; \\
\mathrm{Orb}(x; y, z) &\overset{\text{def}}{\longleftrightarrow} \mathrm{Init}(x) \wedge \Big(\bigvee_{i \leq \kappa} \mathrm{d}_i(x, y)\Big) \wedge \Big(\bigvee_{j \leq \kappa} \mathrm{d}_j(x, z)\Big);
\end{aligned}$$

and for $m \in M$,

$$\Gamma_m(y, z) \overset{\text{def}}{\longleftrightarrow} \exists x \Big(\mathrm{Orb}(x; y, z) \wedge \bigwedge_{0 < i \leq l} [\mathrm{Tp}_i(x) \to \Phi_m^{f_i}(x; y, z)]\Big).$$

Claim. For $b \in A'$:

(i) $\mathbf{A}' \models \mathrm{Un}(b)$ if and only if $b \in A$;

(ii) $\mathbf{A}' \models \mathrm{Flag}(b)$ if and only if $b = \langle a, j\rangle$ for some $a \in I$ and $j \leq \mathrm{typ}(a)$;

(iii) $\mathbf{A}' \models \mathrm{Init}(b)$ if and only if $b \in I$;

(iv) $\mathbf{A}' \models \mathrm{Tp}_i(b)$ if and only if $b \in I$ and $\mathrm{typ}(b) = i$;

(v) For a, b, $c \in A'$, $\mathbf{A}' \models \mathrm{Orb}(a; b, c)$ if and only if $a \in I$ and b, $c \in O_a$;

(vi) For $m \in M$, and b, $c \in A'$, $\mathbf{A}' \models \Gamma_m(b, c)$ if and only if b, $c \in A$, and $\mathbf{A} \models m(b) = c$.

The first four claims follow immediately from our construction of $\mathbf{A}'$.

For (v), if $\mathbf{A}' \models \mathrm{Orb}(a;b,c)$, then $a \in I$, and $\mathbf{A}' \models \mathrm{d}_i(a,b) \wedge \mathrm{d}_j(a,c)$ for some i, j and so b, $c \in O_a$. Conversely, if b, $c \in O_a$ then since $f|_{O_a}$ is a coding function for $\mathbf{O}_a$, and a is $f|_{O_a}$-initial, there are i, $j \leq \kappa$ such that $\mathbf{A}' \models \mathrm{d}_i(a,b) \wedge \mathrm{d}_j(a,c)$. Note that since $|M| = \kappa$ then every one-generated member of $\mathcal{V}$ will have cardinality at most κ.

For (vi), let $m \in M$ and b, $c \in A'$. Then

$$\mathbf{A}' \models \Gamma_m(b,c)$$

implies b, $c \in O_a$ for some $a \in I$, with $\mathrm{typ}(a) = i$, and

$$\mathbf{A}' \models \Phi_m^{f_i}(a;b,c).$$

But then

$$\langle O_a,\, f'|_{O_a}\rangle \models \Phi_m^{f_i}(a;b,c),$$

since O_a is a subuniverse of $\mathbf{A}'$ containing a, b and c, and $\Phi_m^{f_i}$ is preserved under subuniverses. This is equivalent to

$$\mathbf{O}_a \models m(b) = c$$

by Lemma 12.13 since a is $f'|_{O_a}$-initial, and $\mathrm{typ}(a) = i$. Thus

$$\mathbf{A} \models m(b) = c.$$

The converse of (vi) is handled similarly.

Thus we have interpreted $\mathbf{A}$ into $\mathbf{A}'$, and it follows that we can interpret any algebra of $\mathcal{V}$ into some mono-unary algebra, using the formulas given above. Therefore since $\mathcal{V}$ is finitely axiomatizable, it is decidable. □

To complete the proof of Theorem 12.2, we will show that if $\mathbf{M}$ is non-linear, then $\mathcal{V}$ is both unstructured and ω-unstructured. Lemma 12.7 demonstrated that linearity can fail for two distinct reasons; either,

(1) $G_u \neq G$ for some $u \in M$ or,

(2) $\mathbf{M}$ is not quasi-linear.

In either case, we will establish that $\mathcal{V}$ is both unstructured and ω-unstructured by interpreting the theory of two equivalence relations into $\mathcal{V}$, with an interpretation that simultaneously interprets the class of finite sets with two equivalence relations into the class of finite algebras in $\mathcal{V}$. It is not too difficult to show that the class of sets with two equivalence relations is both unstructured and ω-unstructured, and so it will follow that $\mathcal{V}$ also possesses these properties.

LEMMA 12.15 *If* $\mathbf{M}$ *is not quasi-linear, but* $G_x = G$ *for all* $x \in M$*, then* $\mathcal{V}$ *is both unstructured and* ω*-unstructured.*

PROOF. Since $\mathbf{M}$ is not quasi-linear, then for some u, $v \in M$, we have that u and v are not $\propto$-comparable. It follows that u, $v \notin G$, and that the variety $\mathcal{V}$ cannot satisfy any equation of the form

$$ug(x) \approx wv(x)$$

or

$$vg(x) \approx wu(x)$$

for any $w \in M$, and $g \in G$.

The basic idea of the proof is that since u and v are not $\propto$-comparable, then as unary operations acting on $\mathcal{V}$-free algebras, they are somewhat unrelated, and so the two equivalence relations defined by the kernels of these operations are also unrelated. This allows us to semantically embed the undecidable theory of two equivalence relations into $\mathcal{V}$.

Let the structure $\mathbf{A} = \langle A, E_1, E_2\rangle$ be a set with two equivalence relations E_1 and E_2 on it, and let $\mathbf{F} = \mathbf{F}_{\mathcal{V}}(A)$ be the $\mathcal{V}$-free algebra freely generated by the set A. Consider the sets

$$S_1 = \{\langle u(a), u(b)\rangle \,:\, a,\, b \in A \text{ and } \langle a, b\rangle \in E_1\}$$

$$S_2 = \{\langle v(a), v(b)\rangle \,:\, a,\, b \in A \text{ and } \langle a, b\rangle \in E_2\}$$

and the congruence Θ on $\mathbf{F}$ generated by $S_1 \cup S_2$.

Claim 1. For $a \in A$ and $g \in G$, we have

(i) $g(a)/\Theta = \{g(a)\}$;

(ii) $ug(a)/\Theta = \{ug(b) \,:\, b \in A \text{ and } \langle a, b\rangle \in E_1\}$;

(iii) $vg(a)/\Theta = \{vg(b) \,:\, b \in A \text{ and } \langle a, b\rangle \in E_2\}$.

To prove (i), note that if $|g(a)/\Theta| > 1$, then since $S_1 \cup S_2$ generates Θ, we can find some $b \in A$, and $w \in M$ such that $g(a) = w(u(b))$ or $g(a) = w(v(b))$. From this it follows that $g = wu$ or $g = wv$ in $\mathbf{M}$, and so either $u \in G$ or $v \in G$, a contradiction.

To prove (ii), note that since $G_u = G$, then for $g \in G$, there is some $w \in M$ such that $ug = wu$. From this it follows that

$$B = \{ug(b) \,:\, b \in A \text{ and } \langle a, b\rangle \in E_1\} \subseteq ug(a)/\Theta$$

since for $\langle a, b\rangle \in E_1$,

$$\langle ug(a), ug(b)\rangle = \langle w(u(a)), w(u(b))\rangle \text{ and } \langle u(a), u(b)\rangle \in S_1.$$

Thus to establish equality, it will suffice to show that B is a block of the congruence Θ. To do this, it will be enough to show that if $\langle \mu, \nu\rangle \in S_1 \cup S_2$, $w \in M$, and $w(\mu) \in B$, then $w(\nu) \in B$.

First of all, if $\langle \mu, \nu\rangle \in S_2$, say $\langle \mu, \nu\rangle = \langle v(b), v(c)\rangle$, and $w \in M$, then it is impossible for $wv(b)$ to be equal to $ug(d)$ for any $d \in A$, for this would imply that

$$\mathcal{V} \models ug(x) \approx wv(x),$$

forcing $ug = wv$ in $\mathbf{M}$, and so $u \propto v$, which is a contradiction.

Now suppose that $\langle \mu, \nu\rangle \in S_1$, say $\langle \mu, \nu\rangle = \langle u(b), u(c)\rangle$, and $w \in M$ with $wu(b) = ug(d)$ for some $d \in a/E_1$. If $b = d$, then

$$\mathcal{V} \models wu(x) \approx ug(x),$$

implying

$$wu(c) = ug(c) \in B,$$

since $\langle a, c\rangle \in E_1$.

On the other hand, if $b \neq d$, then

$$\mathcal{V} \models wu(x) \approx ug(y),$$

implying

$$\mathcal{V} \models wu(x) \approx u(y),$$

since g is invertible. But then

$$\mathcal{V} \models u(x) \approx u(y),$$

implying $u = uv$ in $\mathbf{M}$, and so $u \propto v$, which is a contradiction. Therefore $B = ug(a)/\Theta$.

The proof of (iii) is similar to the one given for (ii).

Let $\mathbf{A}^* = \mathbf{F}/\Theta \in \mathcal{V}$, and consider the following formulas. Let

$$\text{Un}(x) \stackrel{\text{def}}{\longleftrightarrow} \forall y\Big(\bigwedge_{m \in M-G} m(y) \not\approx x\Big);$$

$$\text{Eq}(x, y) \stackrel{\text{def}}{\longleftrightarrow} \bigvee_{g \in G} g(x) \approx y;$$

$$\text{E}_1(x, y) \stackrel{\text{def}}{\longleftrightarrow} \text{Un}(x) \wedge \text{Un}(y) \wedge \exists x', y' \big(\text{Eq}(x, x') \wedge \text{Eq}(y, y') \wedge u(x') \approx u(y')\big);$$

and

$$\mathrm{E}_2(x,y) \quad \overset{\text{def}}{\longleftrightarrow} \quad \mathrm{Un}(x) \wedge \mathrm{Un}(y) \wedge \exists x', y' \big(\mathrm{Eq}(x,x') \\ \wedge \mathrm{Eq}(y,y') \wedge v(x') \approx v(y')\big).$$

We shall show that these formulas provide a way to interpret $\mathbf{A}$ into $\mathbf{A}^*$. The following claim establishes this.

Claim 2.

(i) For $\mu \in A^*$, $\mathbf{A}^* \models \mathrm{Un}(\mu)$ if and only if $\mu = g(a)/\Theta$ for some $a \in A$.

(ii) $\mathrm{Eq}(x,y)$ defines an equivalence relation on $\mathbf{A}^*$ such that if μ, $\nu \in A^*$, then

$$\mathbf{A}^* \models \mathrm{Un}(\mu) \wedge \mathrm{Un}(\nu) \wedge \mathrm{Eq}(\mu,\nu)$$

if and only if

$$\mu = g(a)/\Theta \text{ and } \nu = h(a)/\Theta$$

for some $a \in A$, and g, $h \in G$.

(iii) For $i = 1$, 2 and μ, $\nu \in A^*$,

$$\mathbf{A}^* \models \mathrm{E}_i(\mu,\nu)$$

if and only if

$$\mu = g(a)/\Theta \text{ and } \nu = h(b)/\Theta$$

for some h, $g \in G$ and a, $b \in A$ with $\langle a, b\rangle \in E_i$.

To see that (i) holds, let $\mu = g(a)/\Theta$, and $m(w(b)/\Theta) = \mu$ for some $m \in M - G$. Then

$$\langle mw(b), g(a)\rangle \in \Theta$$

implying $mw(b) = g(a)$ in $\mathbf{F}$ and so $mw = g$ in $\mathbf{M}$ contradicting $m \notin G$. Thus

$$\mathbf{A}^* \models \mathrm{Un}(\mu), \text{ if } \mu = g(a)/\Theta.$$

Conversely, if $\mu = w(a)/\Theta$ for some $w \in M - G$, and $a \in A$, then clearly $\mathbf{A}^* \models \neg\mathrm{Un}(\mu)$.

To prove (ii), note that since $\mathbf{G}$ is a group, $\mathrm{Eq}(x,y)$ defines an equivalence relation on $\mathbf{A}^*$. Let $\mu = g(a)/\Theta$ and $\nu = h(b)/\Theta$. Then

$$\mathbf{A}^* \models \mathrm{Eq}(\mu,\nu)$$

if and only if

$$g(a)/\Theta = kh(b)/\Theta \text{ for some } k \in G$$

if and only if

$$a = b \text{ in } A$$

since $g(a)/\Theta = \{g(a)\}$.

To prove (iii), let $i = 1$, and $\mu = g(a)/\Theta$, $\nu = h(b)/\Theta$ for some g, $h \in G$, and a, $b \in A$ with $\langle a, b\rangle \in E_1$. Setting $\mu' = a/\Theta$, and $\nu' = b/\Theta$, we see that

$$\mathbf{A}^* \models \mathrm{Eq}(\mu, \mu') \wedge \mathrm{Eq}(\nu, \nu') \wedge u(\mu') = u(\nu'),$$

since $\langle u(a), u(b)\rangle \in \Theta$. Thus $\mathbf{A}^* \models \mathrm{E}_1(\mu, \nu)$.

Conversely, if

$$\mathbf{A}^* \models \mathrm{E}_1(\mu, \nu),$$

then

$$\mu = g(a)/\Theta \text{ and } \nu = h(b)/\Theta$$

for some g, $h \in G$ and a, $b \in A$; and

$$ug'(a)/\Theta = uh'(b)/\Theta$$

for some g', $h' \in G$.

But then it follows from the previous claim that

$$ug' = uh' \text{ in } \mathbf{M}, \text{ and } \langle a, b\rangle \in E_1.$$

The case $i = 2$ is identical.

Thus the structures

$$\mathbf{A} = \langle A, E_1, E_2\rangle \text{ and } \langle \mathrm{Un}^{\mathbf{A}^*}, \mathrm{E}_1{}^{\mathbf{A}^*}, \mathrm{E}_2{}^{\mathbf{A}^*}\rangle/\mathrm{Eq}^{\mathbf{A}^*}$$

are isomorphic via the map which sends a in A to the Eq class in $\mathbf{A}^*$ that contains a/Θ. Since the structure $\mathbf{A}$ was arbitrary, we have provided a scheme for interpreting the theory of two equivalence relations into $\mathcal{V}$. Noting that if the set A is finite, then so is A^*, it follows that $\mathcal{V}$ is both unstructured and ω-unstructured. □

We now present the final lemma needed to prove Theorem 12.2.

LEMMA 12.16 *If for some $u \in M$, $G_u \neq G$, then $\mathcal{V}$ is both unstructured and ω-unstructured.*

PROOF. The proof is similar to, and more subtle than, the proof of the previous lemma. As in that proof, we will interpret the theory of two equivalence relations into $\mathcal{V}$.

The hypothesis implies that **M** is non-linear, and also that $G - G_u$ is nonempty. A useful observation is that

$$\mathcal{V} \models u(x) \not\approx u(y); \tag{12.1}$$

for, otherwise, we would have $u = um$ for all $m \in M$, and in particular, $u \leq ug$ for all $g \in G - G_u$, a contradiction.

Let $\mathbf{A} = \langle A, E_1, E_2 \rangle$ be a nonempty set with two equivalence relations, E_1 and E_2. Let A' be a set disjoint from, but in bijective correspondence with, A via the map $a \mapsto a'$, and let $\mathbf{F} = \mathbf{F}_{\mathcal{V}}(A \cup A')$. Consider the sets

$$\begin{aligned}
S_1 &= \{\langle ug(a), ug(b)\rangle : a, b \in A, \ g \in G_u \text{ and } \langle a, b\rangle \in E_1\}; \\
S_2 &= \{\langle ug(a), ug(b)\rangle : a, b \in A, \ g \in G - G_u \text{ and } \langle a, b\rangle \in E_2\}; \\
S_3 &= \{\langle g(a'), h(a')\rangle : a \in A \text{ and } g, h \in G\}; \\
S_4 &= \{\langle ug(a'), ug(a)\rangle : a \in A \text{ and } g \in G - G_u\}
\end{aligned}$$

and the congruence Θ on $\mathbf{F}$ generated by $S_1 \cup S_2 \cup S_3 \cup S_4$.

Claim 1. For $a \in A$ and $g \in G$,

(i) $g(a)/\Theta = \{g(a)\}$;

(ii) $g(a')/\Theta = \{h(a') : h \in G\}$;

(iii) If $g \in G_u$, then $ug(a)/\Theta = \{ug(b) : b \in A \text{ and } \langle a, b\rangle \in E_1\}$;

(iv) If $g \in G - G_u$, then

$$\begin{aligned}
ug(a)/\Theta = \ & \{uh(b) : b \in A, \ h \in G - G_u \text{ and } \langle a, b\rangle \in E_2\} \\
& \cup \{uk(b') : b \in A, \ k \in G \text{ and } \langle a, b\rangle \in E_2\}.
\end{aligned}$$

The proof of this claim is similar to one from the previous lemma. A complete proof of (iii) will be given in order to illustrate the techniques needed to prove this claim.

It is clear that if $g \in G_u$ and $a \in A$, then

$$B = \{ug(b) : b \in A \text{ and } \langle a, b\rangle \in E_1\} \subseteq ug(a)/\Theta.$$

In order to establish equality, it suffices to show that B is a block of Θ. To do this it will be enough to show that if

$$\langle \mu, \nu\rangle \in S_1 \cup S_2 \cup S_3 \cup S_4 \cup S_4^{\cup},$$

$w \in M$, and $w(\mu) \in B$, then so is $w(\nu)$. A case by case study is required.

Let $\langle \mu, \nu \rangle = \langle uh(b), uh(c) \rangle \in S_1$, with $h \in G_u$ and $\langle b, c \rangle \in E_1$. If

$$w(\mu) = ug(d) \in B \quad \text{for } w \in M,$$

then

$$wuh(b) = ug(d) \ \text{ in } \mathbf{F}.$$

Now, $b \neq d$ would force $u(x) = u(y)$ to hold in $\mathcal{V}$, since $\mathbf{F}$ is $\mathcal{V}$-free, and b and d are generators. This is contrary to (12.1), and so b must equal d.

So $\mathcal{V} \models wuh(x) \approx ug(x)$, and thus

$$w(\nu) = wuh(c) = ug(c) \in B$$

since $b = d$ implies $\langle a, c \rangle \in E_1$.

Let $\langle \mu, \nu \rangle = \langle uh(c), uh(d) \rangle \in S_2 \cup S_4 \cup S_4^{\cup}$, with $h \in G - G_u$ and c, $d \in A \cup A'$. If $w(\mu) = ug(d) \in B$ for $w \in M$, then

$$\mathcal{V} \models wuh(x) \approx ug(x),$$

and so

$$wuh = ug \ \text{ in } \mathbf{M}.$$

Then $u \leq uhg^{-1}$ implying $hg^{-1} \in G_u$. But then $h \in G_u$, since $g \in G_u$, contrary to $h \in G - G_u$.

Finally, let $\langle \mu, \nu \rangle = \langle h(b'), k(b') \rangle \in S_3$ with $b \in A$ and h, $k \in G$. If

$$w(\mu) = ug(d) \in B,$$

then

$$\mathcal{V} \models wh(x) \approx ug(y),$$

since $b' \neq d$ are generators of $\mathbf{F}$. This implies, since g is invertible, that

$$\mathcal{V} \models u(x) \approx u(y), \quad \text{contrary to (12.1).}$$

Thus we have established that B is a block of Θ, and so have shown that $ug(a)/\Theta = B$.

Let $\mathbf{A}^* = \mathbf{F}/\Theta$ in $\mathcal{V}$, and consider the following formulas. Let

$$\mathrm{Gen}(x) \overset{\text{def}}{\longleftrightarrow} \forall y \Big(\bigwedge_{m \in M - G} m(y) \not\approx x \Big);$$

$$\mathrm{Eq}(x, y) \overset{\text{def}}{\longleftrightarrow} \bigvee_{g \in G} g(x) \approx y;$$

$$\mathrm{Un}(x) \overset{\text{def}}{\longleftrightarrow} \mathrm{Gen}(x) \wedge \exists y \, (\mathrm{Eq}(x, y) \wedge x \not\approx y);$$

$$\begin{aligned}
\mathrm{G}_u(x) &\overset{\text{def}}{\longleftrightarrow} \mathrm{Un}(x) \wedge \forall y([\mathrm{Gen}(y) \wedge \neg\mathrm{Un}(y)] \to u(x) \not\approx u(y))\,; \\
\mathrm{E}_1(x,y) &\overset{\text{def}}{\longleftrightarrow} \mathrm{Un}(x) \wedge \mathrm{Un}(y) \wedge \exists x', y'(\mathrm{Eq}(x,x') \wedge \mathrm{Eq}(y,y') \wedge \\
&\qquad \mathrm{G}_u(x') \wedge \mathrm{G}_u(y') \wedge u(x') \approx u(y'));
\end{aligned}$$

and

$$\begin{aligned}
\mathrm{E}_2(x,y) &\overset{\text{def}}{\longleftrightarrow} \mathrm{Un}(x) \wedge \mathrm{Un}(y) \wedge \exists x', y'(\mathrm{Eq}(x,x') \wedge \mathrm{Eq}(y,y') \wedge \\
&\qquad \neg\mathrm{G}_u(x') \wedge \neg\mathrm{G}_u(y') \wedge u(x') \approx u(y')).
\end{aligned}$$

Claim 2. Let μ, $\nu \in A^*$.

(i) $\mathbf{A}^* \models \mathrm{Gen}(\mu)$ if and only if $\mu = g(a)/\Theta$ or $g(a')/\Theta$ for some $a \in A$ and $g \in G$.

(ii) $\mathrm{Eq}(x, y)$ defines an equivalence relation on $\mathbf{A}^*$.

(iii) $\mathbf{A}^* \models \mathrm{Un}(\mu)$ if and only if $\mu = g(a)/\Theta$ for some $a \in A$ and $g \in G$.

(iv) $\mathbf{A}^* \models \mathrm{Un}(\mu) \wedge \mathrm{Un}(\nu) \wedge \mathrm{Eq}(\mu, \nu)$ if and only if $\mu = g(a)/\Theta$ and $\nu = h(a)/\Theta$ for some $a \in A$ and g, $h \in G$.

(v) $\mathbf{A}^* \models \mathrm{G}_u(\mu)$ if and only if $\mu = h(a)/\Theta$ for some $a \in A$ and $h \in G_u$.

(vi) For $i = 1$, 2, $\mathbf{A}^* \models \mathrm{E}_i(\mu, \nu)$ if and only if $\mu = g(a)/\Theta$, $\nu = h(b)/\Theta$ for some g, $h \in G$ and a, $b \in A$ with $\langle a, b\rangle \in E_i$.

The proof of (i) is similar to the proof of assertion (i) of Claim 2 in Lemma 12.15.

Assertion (ii) follows from the fact that $\mathbf{G}$ is a group.

To prove (iii), let $\mathbf{A}^* \models \mathrm{Un}(\mu)$. Then $\mu = g(a)/\Theta$ or $g(a')/\Theta$ for some $a \in A$ and $g \in G$. If $\mu = g(a')/\Theta$, then for all $\lambda \in A^*$ with $\mathbf{A}^* \models \mathrm{Eq}(\mu, \lambda)$ we have $\lambda = hg(a')/\Theta$ for some $h \in G$. But then $\mu = \lambda$ in $\mathbf{A}^*$, since $\langle g(a'), hg(a')\rangle \in \Theta$. Thus $\mathbf{A}^* \models \mathrm{Un}(\mu)$ forces $\mu = g(a)/\Theta$. The converse is easier, and follows from Claim 1.

Statement (iv) follows from (ii) and (iii).

To prove (v), let $\mathbf{A}^* \models \mathrm{G}_u(\mu)$. Then $\mu = g(a)/\Theta$ for some $a \in A$ and $g \in G$. If $g \notin G_u$, then setting $\lambda = a'/\Theta$ we have

$$\mathbf{A}^* \models \mathrm{Gen}(\lambda) \wedge \neg\mathrm{Un}(\lambda) \wedge u(\mu) = u(\lambda),$$

since $\langle ug(a), u(a')\rangle \in \Theta$ when $g \in G - G_u$ and $a \in A$. But this is contrary to $\mathbf{A}^* \models \mathrm{G}_u(\mu)$, and therefore $g \in G_u$. The converse is similar, and uses the facts about Θ proved in Claim 1.

To prove (vi), suppose that $\mathbf{A}^* \models \mathrm{E}_1(\mu,\nu)$. Then $\mu = g(a)/\Theta$, $\nu = h(b)/\Theta$ for some h, $g \in G$ and a, $b \in A$, and for some g', $h' \in G_u$, $ug'(a)/\Theta = uh'(b)/\Theta$ in $\mathbf{A}^*$. But then $\langle ug'(a), uh'(b)\rangle \in \Theta$, and so by Claim 1, we conclude that $\langle a, b\rangle \in E_1$. The converse is easier, and the claim for $\mathrm{E}_2(x,y)$ is handled similarly.

Thus the structures

$$\mathbf{A} = \langle A, E_1, E_2\rangle \quad \text{and} \quad \langle \mathrm{Un}^{\mathbf{A}^*}, \mathrm{E}_1{}^{\mathbf{A}^*}, \mathrm{E}_2{}^{\mathbf{A}^*}\rangle/\mathrm{Eq}^{\mathbf{A}^*}$$

are isomorphic via the map which sends a in A to the Eq class in $\mathbf{A}^*$ that contains a/Θ. As in Lemma 12.15 we conclude that $\mathcal{V}$ is both unstructured and ω-unstructured. □

Proof of Theorem 12.2. Using the Lemmas 12.14, 12.15 and 12.16, we can immediately conclude that if $\mathcal{V}$ is 1-sorted, then it is decidable if and only if it is linear, and if undecidable then it is both unstructured and ω-unstructured. For the multi-sorted case several modifications of the 1-sorted proof are required.

We can associate a partial semigroup $\mathbf{M}(\mathcal{V})$ to any k-sorted unary variety $\mathcal{V}$ in the following way. Let $\mathbf{M}(\mathcal{V})$ have universe $F_{\mathcal{V}}(\langle \mathbf{x}_1, \ldots, \mathbf{x}_k\rangle)$ and define $m(\mathbf{x}_i)\circ n(\mathbf{x}_j)$ in $\mathbf{M}(\mathcal{V})$ to be $m(n(\mathbf{x}_j))$ when the sort of n is equal to i. As in the 1-sorted case we will often regard the elements of $M(\mathcal{V})$ as terms in the language of $\mathcal{V}$.

Within $M(\mathcal{V})$ there are k subgroups $\mathbf{G}_1$, $\mathbf{G}_2$, …, $\mathbf{G}_k$ defined by

$$G_i = \{g(\mathbf{x}_i) \in M(\mathcal{V}) \,:\, \text{the sort of } g \text{ is } i \text{ and } g \text{ is left and right invertible}\}.$$

By extending Definition 12.4 to the structure $\mathbf{M}(\mathcal{V})$ in the natural way and defining, for $m(\mathbf{x}_i) \in M(\mathcal{V})$,

$$G_m = \{g \in G_i \,:\, m \leq mg\}.$$

we have that $\mathcal{V}$ is linear if and only if the structure $\mathbf{M}(\mathcal{V})$ is quasi-linear and $G_m = G_i$ for all $m = m(\mathbf{x}_i)$ in $M(\mathcal{V})$.

In the k-sorted linear case, the definition of the orbit of an element a in an algebra $\mathbf{A}$ should be changed in the following manner. For $\mathbf{A}$ in $\mathcal{V}$, let $C(\mathbf{A})$ be the set of elements of A that are in the range of $t^{\mathbf{A}}(x)$ for some constant term t. Then given a in A, let

$$O_a = \{m^{\mathbf{A}}(a) \,:\, m(x) \text{ is a term and } \sigma(x) = \sigma(a)\} \cup C(\mathbf{A}).$$

Under the assumption of linearity we conclude, as in Lemma 12.9, that if a is in A then the set of subuniverses of $\mathbf{A}$ contained in O_a and containing

$C(\mathbf{A})$ is linearly ordered under inclusion and that every such subuniverse is equal to O_b for some b. With this modification the proof of decidability proceeds as in the 1-sorted case.

With the modified notions of linearity and quasi-linearity simple variants of Lemmas 12.15 and 12.16 can be used to show that if $\mathcal{V}$ is not linear then it is both unstructured and ω-unstructured.

□

COROLLARY 12.17 *Let $\mathcal{V}$ be a locally finite, strongly Abelian variety of finite type. Then the following are equivalent:*

(i) *$\mathcal{V}$ is undecidable.*

(ii) *$\mathcal{V}$ is hereditarily undecidable.*

(iii) *The class of finite members of $\mathcal{V}$ is hereditarily undecidable.*

(iv) *The class of finite members of $\mathcal{V}$ is undecidable.*

(v) *$\mathcal{V}$ is ω-unstructured.*

(vi) *$\mathcal{V}$ is unstructured.*

PROOF. We have often remarked that each of (v) and (vi) implies (ii) and that (ii) implies (i), for any class $\mathcal{V}$. It is also obvious that (iii) implies (ii) and (iv). On the other hand, if a locally finite, strongly Abelian variety $\mathcal{V}$ satisfies (i), then either (v) and (vi) hold via Theorem 11.9, or $\mathcal{V}[d]$ is essentially unary. In the latter case, $\mathcal{V}[d]$ is undecidable by Lemma 11.8, hence is non-linear by Theorem 12.2, whence is unstructured and ω-unstructured by Lemma 12.15 or 12.16. Then it follows that $\mathcal{V}$ satisfies (v) and (vi), by Lemma 11.8. Thus (i) implies (v) and (vi).

Since the class of finite graphs is hereditarily undecidable then (v) implies (iii). Finally, if $\mathcal{V}$ is decidable then we have shown that it is interpretable into the class of all mono-unary algebras and that the finite members of $\mathcal{V}$ are interpretable into the class of finite mono-unary algebras. By Rabin [1969] we know that the class of finite mono-unary algebras is decidable and so it follows that the class of finite members of $\mathcal{V}$ is also decidable. Thus (iv) implies (i). □

COROLLARY 12.18 *Let $\mathcal{V}$ be a locally finite, strongly Abelian variety of finite type.*

(i) *If $\mathcal{V}$ is decidable, then $\mathcal{V}$ is interpretable into the class of algebras with one unary operation.*

(ii) *If $\mathcal{V}$ is undecidable, then the class of algebras with two unary operations is interpretable into $\mathcal{V}$.*

Proof. This follows since the theory of two functions is interpretable into the theory of two equivalence relations as well as into the theory of graphs, and interpretability is a transitive relation. □

Combining the results from Chapters 7, 11 and 12, we arrive at the following Theorem.

THEOREM 12.19 *Let $\mathcal{V}$ be a locally finite, locally strongly solvable variety of finite type. Then $\mathcal{V}$ is a decidable variety if and only if*

(i) *$\mathcal{V}$ is strongly Abelian with essential arity k, for some finite k,*

(ii) *there is a term $d(x_1, \ldots, x_k)$ such that $d^{\mathbf{F}_{\mathcal{V}}(\mathbf{x},\mathbf{y})}$ is essentially k-ary and δ-injective, and*

(iii) *the associated essentially unary k-sorted variety $\mathcal{V}[d]$ is linear.*

Moreover, there is an algorithm to determine precisely which finite strongly solvable algebras of finite type generate decidable varieties.

Proof. Let the hypotheses hold, and assume that $\mathcal{V}$ is decidable. Then by Corollaries 3.2 and 7.2, $\mathcal{V}$ is strongly Abelian. Theorem 0.17 and its proof (especially the proof of part (iv)) demonstrate that $\mathcal{V}$ is finitely generated and that the finite generating algebra has essential arity k for some finite k. Theorem 0.17 also shows that this generating algebra has an essentially k-ary δ-injective term operation if and only if $\mathbf{F}_{\mathcal{V}}(\mathbf{x}, \mathbf{y})$ does. Then we have that (ii) holds by Lemma 11.8 and Theorem 11.9. Theorem 12.2 gives us (iii). Conversely, if (i)–(iii) hold, then the decidability of $\mathcal{V}$ follows from Lemma 11.8 and Lemma 12.14.

Using the structure theory that we have just developed, along with the fact that we can effectively bound the essential arity of any finite strongly Abelian algebra (Theorem 0.17), an algorithm can be devised to determine if a finite strongly solvable algebra of finite type conforms to the structure required to guarantee that the algebra generates a decidable variety. □

Part III

The decomposition

Chapter 13

The decomposition theorem

In this chapter, we assume that $\mathcal{V}$ is a structured locally finite variety. It follows from the work of Parts I and II that $\mathcal{V}$ is the join of a strongly Abelian variety $\mathcal{V}_1$, an affine variety $\mathcal{V}_2$, and a discriminator variety $\mathcal{V}_3$. (See Definition 1.1 and Theorems 4.1, 5.4 and 9.6.) In this chapter, we shall prove that $\mathcal{V}$ is the product of these three varieties. There are several equivalent ways to formulate the result (see Theorem 0.5):

(1) Every subdirect product $\mathbf{C} \leq \mathbf{C}_1 \times \mathbf{C}_2 \times \mathbf{C}_3$ with $\mathbf{C}_i \in \mathcal{V}_i$ (for $1 \leq i \leq 3$) is direct, i.e., $\mathbf{C} = \mathbf{C}_1 \times \mathbf{C}_2 \times \mathbf{C}_3$.

(2) Every subdirect product $\mathbf{C} \leq \mathbf{C}_1 \times \mathbf{C}_2 \times \mathbf{C}_3$ with $\mathbf{C}_i \in \mathcal{V}_i$ (for $1 \leq i \leq 3$) and $\mathbf{C}_i$ finite is direct.

(3) If $\mathbf{C} = \mathbf{F}_{\mathcal{V}}(3)$ and $\mathbf{C}_i = \mathbf{F}_{\mathcal{V}_i}(3)$, then $\mathbf{C} \cong \mathbf{C}_1 \times \mathbf{C}_2 \times \mathbf{C}_3$.

(4) There exists a term $t(x_1, x_2, x_3)$ such that $t(x_1, x_2, x_3) \approx x_i$ is an identity of $\mathcal{V}_i$ (for $1 \leq i \leq 3$); i.e., the triple of varieties $\langle \mathcal{V}_1, \mathcal{V}_2, \mathcal{V}_3 \rangle$ is independent.

The concepts defined below are key ingredients in our proof. Using them, we prove that the pair $\langle \mathcal{V}_2, \mathcal{V}_3 \rangle$ is independent. Next we prove that if $\langle \mathcal{V}_1, \mathcal{V}_2, \mathcal{V}_3 \rangle$ is not independent, then there exists a "strange pair" of varieties. Finally, we prove that no strange pair can exist.

Definition 13.1 Let $\mathbf{Q} \leq \mathbf{A}_0 \times \mathbf{A}_1$ be a subdirect product with projection homomorphisms $p_0 : \mathbf{Q} \to \mathbf{A}_0$ and $p_1 : \mathbf{Q} \to \mathbf{A}_1$. For $f \in \mathrm{Pol}_n \mathbf{Q}$ we use $p_i(f)$ ($i \in \{0,1\}$) to denote the polynomials operations of $\mathbf{A}_i$ such that

$f(\bar{x}) = \langle p_0(f)(p_0\bar{x}), p_1(f)(p_1\bar{x})\rangle$ for all $\bar{x} \in Q^n$. Now let $f \in \mathrm{Pol}_1\mathbf{Q}$ and let $i \in \{0,1\}$. We say that f is **i-transparent** iff $p_i(f)$ is the identity function on A_i; **i-constant** iff $p_i(f)$ is a constant function; an **i-projection** iff f is both i-transparent and $1-i$ -constant. We say that $\mathbf{Q}$ **has polynomial projections** iff there exist a 0-projection $f \in \mathrm{Pol}_1\mathbf{Q}$ and a 1-projection $g \in \mathrm{Pol}_1\mathbf{Q}$.

Recall that $\mathrm{E}(\mathbf{Q})$ denotes the set of all polynomial functions $e \in \mathrm{Pol}_1\mathbf{Q}$ satisfying $e^2 = e$. Thus $\mathbf{Q}$ has polynomial projections if and only if there exists $\langle a,b\rangle \in Q$ such that the functions $e_0(\langle x,y\rangle) = \langle x,b\rangle$ and $e_1(\langle x,y\rangle) = \langle a,y\rangle$ belong to $\mathrm{E}(\mathbf{Q})$.

LEMMA 13.2 *Let $\mathbf{Q} \le \mathbf{A}\times\mathbf{B}$ be a subdirect product with polynomial projections, where $\mathbf{Q}$ is finite. If $\mathbf{Q} \ne \mathbf{A}\times\mathbf{B}$, then the class $\boldsymbol{P}_s(\mathbf{Q})$ is unstructured.*

PROOF. Assume that $\mathbf{Q} \ne \mathbf{A}\times\mathbf{B}$. Choose $\lambda = \langle a,b\rangle \in Q$ and e_0, $e_1 \in \mathrm{E}(\mathbf{Q})$ such that $e_0(\langle x,y\rangle) = \langle x,b\rangle$ and $e_1(\langle x,y\rangle) = \langle a,y\rangle$ for all $\langle x,y\rangle \in Q$. Then put

$$Q_0 = e_0(Q) = \{\langle x,b\rangle : x \in A\};$$

$$Q_1 = e_1(Q) = \{\langle a,y\rangle : y \in B\};$$

$$S(\mu) = \{e_0(\nu) : \nu \in Q \text{ and } e_1(\nu) = e_1(\mu)\} \text{ when } \mu \in Q.$$

For $\mu,\nu \in Q$ we write $\mu \le \nu$ iff $S(\mu) \subseteq S(\nu)$, and $\mu \equiv \nu$ iff $S(\mu) = S(\nu)$. Notice that we have $S(\mu) \subseteq Q_0 = S(\lambda)$, where $\lambda = \langle a,b\rangle$; thus $\mu \le \lambda$ for all $\mu \in Q$.

Our assumption that $\mathbf{Q} \ne \mathbf{A}\times\mathbf{B}$ is equivalent to the existence of some $\mu \in Q$ such that $\mu < \lambda$. Thus we can choose $\tau \in Q$ such that $\tau < \lambda$ and for all $\mu \in Q$, $\tau \le \mu \le \lambda$ implies $\tau \equiv \mu$ or $\mu \equiv \lambda$. Since $S(\tau) = S(e_1(\tau))$, we can assume that $\tau = e_1(\tau)$.

We shall prove that $\boldsymbol{P}_s(\mathbf{Q})$ is an unstructured class by interpreting the class $\mathcal{BP}_1$ of Definition 0.46 into it. Let $\mathbf{C} = \langle C_1, C_0, \le\rangle$ be a member of $\mathcal{BP}_1$ with $C_1 \subseteq \mathcal{P}(X)$. The algebra into which we shall interpret $\mathbf{C}$ is easily defined.

> $\mathbf{D}$ is the subalgebra of $\mathbf{Q}^X$ consisting of all functions $f \in Q^X$ such that $e_i^{(X)}(f) \in Q_i[C_i]$ for $i = 0$ and $i = 1$; i.e., $f \in Q^X$ belongs to $\mathbf{D}$ iff $(e_0 f)^{-1}(\mu) \in C_0$ for all $\mu \in Q_0$ and $(e_1 f)^{-1}(\mu) \in C_1$ for all $\mu \in Q_1$.

Notice that D is simply the intersection of Q^X with the image of $A[C_0] \times B[C_1]$ under the natural identification of $A^X \times B^X$ with $(A \times B)^X$.

$\mathbf{D}$ contains all the constant functions $\hat{c}$ with $c \in Q$. Thus all the polynomial operations of $\mathbf{Q}$, acting coordinatewise in Q^X, are polynomial operations of $\mathbf{D}$. We write e_0 and e_1 for the polynomial functions of $\mathbf{D}$ corresponding to the polynomial functions e_0 and e_1 of $\mathbf{Q}$. We define $S(f)$ for $f \in D$ and the relations $\leq$ and $\equiv$ on $\mathbf{D}$ just as we did for $\mathbf{Q}$ (i.e., using the same first order formulas). Next we define a subset of D and a mapping from this subset to C_1.

$$\Sigma_1 = \{f \in e_1(D) : \hat{\tau} \leq f\};$$

$$C(f) = \{x \in X : f(x) \equiv \lambda\}.$$

It can easily be shown that for $f \in D$, $S(f)$ is identical with the set

$$Q_0[C_0] \cap \prod_{x \in X} S(f(x)).$$

Since C_0 includes all the finite subsets of X, it follows that $f \leq g$ (where $f, g \in D$) iff $f(x) \leq g(x)$ for all $x \in X$. Thus, since τ is maximal $< \lambda$, Σ_1 is just the set of all functions $f \in Q_1[C_1]$ such that $f(x) \equiv \tau$ or $f(x) \equiv \lambda$ for every $x \in X$. Then it follows that

$$f \leq g \text{ iff } C(f) \subseteq C(g) \text{ when } f, g \in \Sigma_1.$$

Moreover, it is easily verified that $C(\Sigma_1) = C_1$. Thus

$$\langle \Sigma_1, \leq \rangle / \equiv \;\; \cong \;\; \langle C_1, \leq \rangle \text{ via the mapping } C.$$

We define a subset Σ_0 of Σ_1.

$$f \in \Sigma_0 \;\longleftrightarrow$$

$$\begin{array}{l} f \in \Sigma_1 \text{ and there exists } g \in e_0(D) \text{ such that} \\ \text{for all } h \in \Sigma_1,\ f \leq h \text{ iff } g \in S(h). \end{array}$$

We shall now show that

$$\Sigma_0 = \{f \in \Sigma_1 : C(f) \in C_0\} \tag{13.1}$$

which will finish the proof—since it implies that $\langle C_1, C_0, \leq \rangle$ is isomorphic to the definable structure $\langle \Sigma_1, \Sigma_0, \leq \rangle / \equiv$.

To prove (13.1), suppose first that $f \in \Sigma_0$, and let $g \in e_0(D)$ be such that $f \leq h$ iff $g \in S(h)$ for all $h \in \Sigma_1$. Recall that $f(x) \equiv \tau$ or $f(x) \equiv \lambda$ for all x. Since $g \in S(f)$ it follows that $g(x) \in S(\tau)$ whenever $f(x) \equiv \tau$. Also, $g(x) \notin S(\tau)$ if $f(x) \equiv \lambda$; for if this were not the case then we could construct $h \in \Sigma_1$ (by changing f at one place x, replacing its value by τ) such that $h < f$ and $g \in S(h)$. Now $C(f) = \{x : f(x) \equiv \lambda\} = g^{-1}(Q_0 - S(\tau))$, and since $g = e_0 g$ it follows from the definition of $\mathbf{D}$ that $C(f) \in C_0$.

Now suppose, conversely, that $f \in \Sigma_1$ and $C(f) \in C_0$. Choose $\beta \in Q_0 - S(\tau)$ and define g to be the function such that $g(x) = \lambda$ if $f(x) \equiv \tau$ and $g(x) = \beta$ if $f(x) \equiv \lambda$. Thus $e_0(g) = g$, $g^{-1}(\beta) = C(f)$, and we have $g \in e_0(D)$. Moreover, it is obvious that $h \in \Sigma_1$ implies $g \in S(h)$ iff $g^{-1}(\beta) \subseteq \{x : h(x) \equiv \lambda\}$, i.e., iff $f \leq h$. This concludes the proof of (13.1) and the proof of this lemma. □

We proceed to prove several lemmas which will allow us to show that certain subdirect products must have polynomial projections. Recall that a ternary operation d on a set U is said to be **Maltsev** iff it obeys the equations $d(x,x,y) \approx y$ and $d(x,y,y) \approx x$.

LEMMA 13.3 *Let $\mathbf{Q} \leq \mathbf{A}_0 \times \mathbf{A}_1$ be a subdirect product, $\mathbf{Q}$ finite, and let d be a Maltsev polynomial operation of $\mathbf{A}_1$. Then for every 1-transparent $f \in \mathrm{Pol}_1\mathbf{Q}$, there exists a 1-transparent $e \in E(\mathbf{Q})$ and a ternary polynomial operation h of $\mathbf{Q}$ such that: $p_1(h) = d$; $h(Q^3) = e(Q) \subseteq f(Q)$; and $h|_{e(Q)}$ is Maltsev.*

PROOF. The set of 1-transparent polynomial functions of $\mathbf{Q}$ is closed under composition. Thus there exists an idempotent 1-transparent polynomial function e with $e(Q) \subseteq f(Q)$. (For example, there is $n > 0$ such that f^n has these properties.) We choose a 1-transparent $e \in \mathrm{E}(\mathbf{Q})$ with $e(Q) \subseteq f(Q)$ and such that $e(Q)$ is minimal. (There exists no polynomial function with these properties whose range is a proper subset of the range of e.) Now we define four sets of polynomials.

$$\Gamma_0 = \{h \in \mathrm{Pol}_3\mathbf{Q} \ : \ p_1(h) = d \text{ and } eh = h\};$$

$$\Gamma_1 = \{h \in \Gamma_0 \ : \ h(e(\mu), e(\mu), e(\mu)) = e(\mu) \text{ for all } \mu \in Q\};$$

$$\Gamma_2 = \{h \in \Gamma_1 \ : \ h(e(\mu), e(\mu), e(\nu)) = e(\nu) \text{ for all } \mu, \nu \in Q\};$$

$$\Gamma_3 = \{h \in \Gamma_2 \ : \ h(e(\nu), e(\mu), e(\mu)) = e(\nu) \text{ for all } \mu, \nu \in Q\}.$$

We shall prove this lemma by showing that each of the sets Γ_i is non-empty. Notice that if $h \in \Gamma_3$, then e and h fulfill the requirements of this lemma.

Firstly, since $p_1(\mathbf{Q}) = \mathbf{A}_1$, there exists $h \in \mathrm{Pol}_3\mathbf{Q}$ with $p_1(h) = d$. Then $eh \in \Gamma_0$, since e is 1-transparent.

Now let $h \in \Gamma_0$. Since $p_1(h) = d$, the function $\delta(x) = h(x,x,x)$ is 1-transparent. Let $e' = \delta^n$, $n > 1$, be an idempotent power of δ. Then e' is 1-transparent and $ee' = e'$, from which we conclude that $e'(Q) = e(Q)$, since we chose e to have minimal range. This means that $e'e = e$. Now define $h'(x,y,z) = \delta^{n-1}h(x,y,z)$. Then $h' \in \Gamma_0$, clearly; and also $h'(ex,ex,ex) = \delta^n(ex) = e'e(x) = e(x)$. Thus we have shown that $\Gamma_1 \neq \emptyset$.

To see that $\Gamma_2 \neq \emptyset$, let $h \in \Gamma_1$. Define $\tau(x,y) = h(x,x,y)$, and then $\tau^{(1)}(x,y) = \tau(x,y)$ and, inductively, $\tau^{(n+1)}(x,y) = \tau(x,\tau^{(n)}(x,y))$. We can choose $n > 0$ such that the operation $t = \tau^{(n)}$ satisfies $t(x,t(x,y)) = t(x,y)$ for all $x,y \in Q$. Observe that $p_1t(x,y) = p_1(y)$ for all $x,y \in Q$ since p_1h is Maltsev. (This càn be proved for all of the operations $\tau^{(k)}$ by induction on k.) Thus for each $a \in Q$, the function $g(x) = t(a,x)$ is a 1-transparent idempotent polynomial function of $\mathbf{Q}$ such that $eg = g$. By the minimality of e, we conclude that $ge = e$. Now define $h'(x,y,z) = \tau^{(n-1)}(x,h(x,y,z))$. Then we have that $h'(ex,ex,ey) = \tau^{(n)}(ex,ey) = ey$ for all $x,y \in Q$, which means that $h' \in \Gamma_2$.

Finally, to see that $\Gamma_3 \neq \emptyset$, let $h \in \Gamma_2$. The construction of the last paragraph, iterating the polynomial operation $\sigma(x,y) = h(x,y,y)$, this time as a function of x with y held fixed, produces (just as above) a polynomial operation $h' \in \Gamma_1$ such that $h'(ex,ey,ey) = ex$. But it is also easy to check that we have $h'(ex,ex,ey) = ey$, since h had this property. Thus Γ_3 is seen to be non-empty, and our proof is complete. □

Recall from Definition 0.21 that if $\mathcal{D}$ is a discriminator variety then it has a discriminator term $d(x,y,z)$. Among other properties, this term obeys Maltsev's equations in $\mathcal{D}$, and also the equation $d(x,y,x) \approx x$.

LEMMA 13.4 *Suppose that $\mathcal{A}$ and $\mathcal{D}$ are varieties of the same type, and that $\mathcal{A}$ is a locally finite Abelian variety, while $\mathcal{D}$ is a discriminator variety. Let $d(x,y,z)$ be any discriminator term for $\mathcal{D}$. There exists a term $t(x,y,z)$ such that $\mathcal{A} \models t(x,y,z) \approx t(y,y,y)$ and $\mathcal{D} \models t(x,y,z) \approx d(x,y,z)$.*

PROOF. We construct a sequence of terms $t^{(k)}$ where $t^{(k)}$ has the variables $x, y_1, \ldots, y_k$. Let

$$t^{(1)}(x,y_1) = d(x,x,y_1)$$

and inductively,

$$t^{(k+1)}(x,y_1,\ldots,y_{k+1}) = d(t^{(k)}(x,y_1,\ldots,y_k),x,y_{k+1}).$$

Now let $s = t^{(n)}$ where $n = |\mathbf{F}_{\mathcal{A}}(2)|$. We define terms s_i, $0 \leq i \leq n$, using the variables u and v. Namely, we put $s_i(u,v) = s(u,\ldots,u,v,\ldots,v)$ in which $x, y_1, \ldots, y_i$ have been replaced by u and $y_{i+1}, \ldots, y_n$ have been replaced by v. Since there are only n distinct binary terms over $\mathcal{A}$, there exist $i < j$ such that $\mathcal{A} \models s_i \approx s_j$. Then since $\mathcal{A}$ is Abelian, we have that

$$\mathcal{A} \models s(x, y_1, \ldots, y_i, u, \ldots, u, y_{j+1}, \ldots, y_n) \approx$$

$$s(x, y_1, \ldots, y_i, v, \ldots, v, y_{j+1}, \ldots, y_n).$$

Now we put

$$t(x, y, z) = s(y, y, \ldots, y, d(x, y, z), \ldots, d(x, y, z), y, \ldots, y),$$

the term obtained from s by replacing the variables $x, y_1, \ldots, y_i$ with y, replacing $y_{i+1}, \ldots, y_j$ with $d(x, y, z)$, and replacing $y_{j+1}, \ldots, y_n$ with y. The above displayed equation implies that $\mathcal{A} \models t(x, y, z) \approx t(y, y, y)$. With a little effort, one also sees that $\mathcal{D} \models t(x, y, z) \approx d(x, y, z)$. □

LEMMA 13.5 *Suppose that $\mathcal{S}$ and $\mathcal{M}$ are varieties of the same type, and that $\mathcal{S}$ is a locally finite, strongly Abelian variety, while $\mathcal{M}$ is Maltsev. Let $d(x, y, z)$ be a Maltsev term for $\mathcal{M}$. There exists a term $t(x, y, z)$ such that $\mathcal{S} \models t(x, y, z) \approx t(y, y, y)$ and $\mathcal{M} \models t(x, y, z) \approx d(x, y, z)$.*

PROOF. The proof is almost identical to the proof of the last lemma. We use the same terms $t^{(k)}(x, y_1, \ldots, y_k)$, $s = t^{(n)}$ where $n = |\mathbf{F}_{\mathcal{S}}(2)|$, and $s_i(u, v)$ $(0 \leq i \leq n)$. Let $\mathcal{S} \models s_i \approx s_j$ with $i < j$. Since $\mathcal{S}$ is strongly Abelian, this implies that $s = t^{(n)}$ is independent of its variable y_j, in $\mathcal{S}$. Then we construct $t(x, y, z)$ from s by replacing $x, y_1, \ldots, y_{j-1}, y_{j+1}, \ldots, y_n$ with y and replacing y_j with $d(x, y, z)$. This term is easily seen to have the required properties. □

LEMMA 13.6 *Let $\mathcal{D}$ be a discriminator variety. Every subdirect product $\mathbf{Q} \leq \mathbf{A}_0 \times \mathbf{A}_1$ where $\mathbf{Q}$ is finite, $\mathbf{A}_0$ is affine and $\mathbf{A}_1 \in \mathcal{D}$, has polynomial projections.*

PROOF. Since $\mathbf{A}_0$ is affine, it has a Maltsev term operation $d_0(x, y, z) = x - y + z$ derived from an Abelian group structure on $\mathbf{A}_0$. We apply Lemma 13.3 and find an idempotent and 0-transparent polynomial function e of $\mathbf{Q}$, and an $h \in \mathrm{Pol}_3\mathbf{Q}$ such that $eh = h$, $p_0(h) = d_0$, and $h|_{e(Q)}$ is Maltsev. This implies that $p_1(h) = d_1$ is Maltsev on the set $p_1e(Q)$. Let $t(x, y, z)$ be the term supplied by Lemma 13.4 (taking $\mathcal{A} = \boldsymbol{V}(\mathbf{A}_0)$). Then choosing some $a = \langle a_0, a_1 \rangle \in e(Q)$, the polynomial functions

$$e_1(x) = t^{\mathbf{Q}}(x, a, a),$$

$$e_0(x) = h(e(x), ee_1e(x), ee_1(a))$$

are a 1-projection and a 0-projection, respectively. In fact, $e_1(\langle x_0, x_1\rangle) = \langle t^{\mathbf{A}_0}(a_0, a_0, a_0), x_1\rangle$, while $e_0(\langle x_0, x_1\rangle) = \langle x_0, a_1\rangle$. □

We are now ready for the first substantial theorem of this chapter.

THEOREM 13.7 $\mathcal{V}_2 \vee \mathcal{V}_3 = \mathcal{V}_2 \otimes \mathcal{V}_3$.

PROOF. By Lemmas 13.2 and 13.6, every subdirect product of a finite algebra in $\mathcal{V}_2$ and a finite algebra in $\mathcal{V}_3$ is direct. This theorem is equivalent to this fact. (See the opening remarks of this chapter.) □

COROLLARY 13.8 (i) *$\mathcal{V}_2 \vee \mathcal{V}_3$ is a Maltsev variety.*

(ii) *There exists a term $t(x, y, z)$ such that t is a discriminator term for $\mathcal{V}_3$ and a Maltsev term for $\mathcal{V}_2$, and $\mathcal{V}_1 \models t(x, y, z) \approx t(y, y, y)$.*

PROOF. The product of two Maltsev varieties is Maltsev. In fact, letting d_2 be a Maltsev term for $\mathcal{V}_2$ and d_3 be a discriminator term for $\mathcal{V}_3$, there exists a term d such that $\mathcal{V}_i \models d \approx d_i$, $i = 2, 3$—since by the preceding theorem, $\mathcal{V}_2$ and $\mathcal{V}_3$ are independent. By a previous lemma, there exists a term t such that $\mathcal{V}_2 \otimes \mathcal{V}_3 \models t \approx d$ and $\mathcal{V}_1 \models t(x, y, z) \approx t(y, y, y)$. □

In order to complete our proof that $\mathcal{V} = \mathcal{V}_1 \otimes \mathcal{V}_2 \otimes \mathcal{V}_3$, we need a very special (and as it turns out, vacuous) concept.

Definition 13.9 Let $\mathcal{S}$ be a locally finite, strongly Abelian variety and $\mathcal{M}$ be a locally finite, Maltsev variety. We say that $\langle \mathcal{S}, \mathcal{M}\rangle$ is a **strange pair** if, in addition, the following hold.

(1) $\mathcal{S} \vee \mathcal{M}$ is structured.

(2) $\mathcal{M}$ is not the variety of one-element algebras.

(3) $\mathcal{S}$ is essentially unary; $\mathbf{F}_{\mathcal{S}}(\mathbf{x})$ has precisely two elements, $\mathbf{x}$ and $\mu(\mathbf{x})$; and $\mathcal{S} \models \mu(x) \approx \mu(y)$. Moreover, for every term $t(x_0, \ldots, x_{n-1})$, either $\mathcal{S} \models t(x_0, \ldots, x_{n-1}) \approx \mu(x_0)$, or there is an $i < n$ such that $\mathcal{S} \models t(x_0, \ldots, x_{n-1}) \approx x_i$. An n-ary term $t(x, \bar{y})$ will be called $\mathcal{S}$-**normal** if $\mathcal{S} \models t(x, \bar{y}) \approx x$, and called $\mathcal{S}$-**constant** if $\mathcal{S} \models t(x, \bar{y}) \approx \mu(x)$.

(4) For every $\mathcal{S}$-normal term $t(x, \bar{y})$, there exists an $\mathcal{S}$-normal term $r(x, \bar{y})$ such that $\mathcal{M} \models r(t(x, \bar{y}), \bar{y}) \approx x$. Moreover, there exists an $\mathcal{S}$-constant term $t(x)$ such that $\mathcal{M} \models t(x) \approx x$.

In the final pages of this chapter, we prove that strange pairs of varieties do not exist. Using that result, we can now establish the main result of this work.

THEOREM 13.10 *Let $\mathcal{V}$ be any structured locally finite variety. There exists a strongly Abelian variety $\mathcal{V}_1$, an affine variety $\mathcal{V}_2$, and a discriminator variety $\mathcal{V}_3$ such that*

$$\mathcal{V} = \mathcal{V}_1 \otimes \mathcal{V}_2 \otimes \mathcal{V}_3.$$

These three subvarieties of $\mathcal{V}$ are uniquely determined.

PROOF. Let $\mathcal{V}$ be a locally finite, structured variety. If $\mathcal{V} = \mathcal{V}_1 \otimes \mathcal{V}_2 \otimes \mathcal{V}_3$ as described, then $\mathcal{V}_1$ can be none other than the class of all strongly Abelian algebras in $\mathcal{V}$; $\mathcal{V}_2$ must be the class of all affine algebras in $\mathcal{V}$; and $\mathcal{V}_3$ can only be the class of all centerless algebras in $\mathcal{V}$. Thus, as we have seen in Parts I and II, $\mathcal{V}_1$, $\mathcal{V}_2$, and $\mathcal{V}_3$ can only be the three subvarieties defined in Definition 1.1. Let us assume that it is not the case that $\mathcal{V} = \mathcal{V}_1 \otimes \mathcal{V}_2 \otimes \mathcal{V}_3$. We shall show that this assumption implies the existence of a strange pair (see Definition 13.9).

Since $\mathcal{V}_2$ and $\mathcal{V}_3$ are independent by Theorem 13.7, we have as an easy corollary of Theorem 0.5 that $\mathcal{V}_1$ and $\mathcal{V}_2 \otimes \mathcal{V}_3$ are not independent. Thus there exists a subdirect product $\mathbf{Q} \leq \mathbf{A}_0 \times \mathbf{A}_1$ which is not direct, where $\mathbf{A}_0 \in \mathcal{V}_1$ and $\mathbf{A}_1 \in \mathcal{V}_2 \otimes \mathcal{V}_3$ and $\mathbf{Q}$ is finite. Lemma 13.5 makes it clear that there is a 1-projection of the form $e(x) = t^{\mathbf{Q}}(a, a, x)$ in $\mathrm{E}(\mathbf{Q})$. By Lemma 13.2, there is no 0-projection.

Recall that we actually have a lot of detailed knowledge of the structure of $\mathbf{A}_0$. According to Lemma 11.3 and Theorem 11.9 (see also the paragraph preceding Lemma 11.4), there exists a term d such that $d^{\mathbf{A}_0}$ is an n-ary decomposition operation (for some n) and $\mathbf{A}_0$ has no polynomial operation depending on more than n variables. Let $a \in Q$ and for $0 \leq i \leq n-1$ let $d^{(i)}(x) = d^{\mathbf{Q}}(a, \ldots, a, x, a, \ldots, a)$, with x in the ith place and the constant a in every other place—this is a polynomial operation of $\mathbf{Q}$. Let $e^{(i)}$ be an idempotent power of $d^{(i)}$. Since $p_0(d^{\mathbf{Q}})$ is the decomposition operation of $\mathbf{A}_0$, it follows that $p_0(e^{(i)}) = p_0(d^{(i)})$ $(= f^{(i)}$ say) where

$$d^{\mathbf{A}_0}(f^{(0)}(x), \ldots, f^{(n-1)}(x)) = x$$

for all x in A_0, and $d^{\mathbf{Q}}(e^{(0)}(x), \ldots, e^{(n-1)}(x))$ is a 0-transparent polynomial operation of $\mathbf{Q}$.

Now we consider the algebras $e^{(i)}(\mathbf{Q}) = \mathbf{Q}\mathrm{I}_{e^{(i)}(Q)}$ induced by $\mathbf{Q}$ on the sets $e^{(i)}(Q)$. (See Definition 0.49 and the second paragraph of §0.6 for this concept.) The algebra $e^{(i)}(\mathbf{Q})$ is a subdirect product of two algebras on the base sets $e_0^{(i)}(A_0)$ and $e_1^{(i)}(A_1)$ where $e_j^{(i)} = p_j(e^{(i)})$. Suppose that for every $i < n$, $e^{(i)}(\mathbf{Q})$ has a 0-projection $p^{(i)}$. Letting $q^{(i)}$ be a polynomial function of $\mathbf{Q}$ which induces $p^{(i)}$, we can easily calculate that

$$q(x) = d^{\mathbf{Q}}(q^{(0)}e^{(0)}(x), \ldots, q^{(n-1)}e^{(n-1)}(x))$$

defines a 0-projection q of $\mathbf{Q}$. Since $\mathbf{Q}$ has no 0-projection, some algebra $e^{(i)}(\mathbf{Q})$, say $e^{(0)}(\mathbf{Q})$, fails to have any 0-projection.

Let us now take a close look at the algebra $e^{(0)}(\mathbf{Q})$, which we shall denote by $\mathbf{B}'$. This algebra is a subdirect product of algebras $\mathbf{S}'$ and $\mathbf{M}'$ that are polynomially equivalent, respectively, to $e_0^{(0)}(\mathbf{A}_0)$ and $e_1^{(0)}(\mathbf{A}_1)$. From the properties of $d^{\mathbf{A}_0}$ we have that $\mathbf{S}'$ has no operations depending on more than one variable. The algebras $\mathbf{S}'$ and $\mathbf{M}'$ are non-trivial since $\mathbf{B}'$ has no 0-projection. The algebra $\mathbf{M}'$ has a Maltsev operation $e_1^{(0)}p(x,y,z)$. (Take for p any Maltsev operation of $\mathbf{A}_1$.)

Next, we choose a 0-transparent $e \in \mathrm{E}(\mathbf{B}')$ having minimal range, and we define $\mathbf{B}$ to be $\mathbf{B}'\mathrm{I}_{e(B')}$, and $\mathbf{S}$ and $\mathbf{M}$ to be the algebras of which $\mathbf{B}$ is the subdirect product. (They are polynomially equivalent, respectively, to $\mathbf{S}'$ and $\mathbf{M}'\mathrm{I}_{e_1(M')}$, where $e_1 = p_1(e)$.)

We have secured the following conditions.

(1) $\mathbf{B} \leq \mathbf{S} \times \mathbf{M}$, a subdirect product.

(2) $\boldsymbol{V}(\mathbf{S})$ is essentially unary, and $\mathbf{S}$ has at least two elements.

(3) $\boldsymbol{V}(\mathbf{M})$ is Maltsev, and $\mathbf{M}$ has at least two elements.

(4) Every 0-transparent polynomial function of $\mathbf{B}$ is a permutation of $\mathbf{B}$.

(5) $\boldsymbol{V}(\mathbf{B}) = \boldsymbol{V}(\mathbf{S}) \vee \boldsymbol{V}(\mathbf{M})$ is structured.

In fact, $\mathbf{S}$ is polynomially equivalent to $\mathbf{S}'$; and (3) and (4) follow from the fact that e is 0-transparent but not a 0-projection of $\mathbf{B}'$, and the fact that e has minimal range: If $\mathbf{B}$ has a 0-transparent polynomial f that is not a permutation, then f is the restriction to B of a polynomial g of $\mathbf{B}'$, and an idempotent power of ege would be 0-transparent and have smaller range than e. Statement (5) is implied by Theorem 0.50 and the discussion preceding that theorem: According to Theorem 0.50, every reduct of $\boldsymbol{V}(\mathbf{B})$ to a finite language is interpretable into a reduct of $\boldsymbol{V}(\mathbf{B}')$ to a finite language, and thence into $\boldsymbol{V}(\mathbf{Q})$, and so finally into $\mathcal{V}$. (These interpretations interpret finite algebras in finite algebras.) Since $\mathcal{V}$ is structured, $\boldsymbol{V}(\mathbf{B})$ must also be.

We consider next the free algebra $\mathbf{U} = \mathbf{F}_{\mathcal{W}}(\mathbf{x})$ generated by one element $\mathbf{x}$, where $\mathcal{W} = \boldsymbol{V}(\mathbf{S})$. Let us call an element $u = s^{\mathbf{U}}(\mathbf{x})$ (where s is a unary term) **invertible** iff there is a term r such that $r^{\mathbf{U}}(s^{\mathbf{U}}(\mathbf{x})) = \mathbf{x}$ in $\mathbf{U}$, i.e., $\boldsymbol{V}(\mathbf{S}) \models r(s(x)) \approx x$. Since $\mathbf{U}$ is finite, an element $s^{\mathbf{U}}(\mathbf{x})$ is invertible iff the term operation $s^{\mathbf{U}}$ is a permutation of U. This makes it easy to see that the relation we now define is a congruence of $\mathbf{U}$: $r^{\mathbf{U}}(\mathbf{x}) \sim s^{\mathbf{U}}(\mathbf{x})$ iff [$r^{\mathbf{U}}(\mathbf{x})$ is invertible iff $s^{\mathbf{U}}(\mathbf{x})$ is invertible]. Now the element $\mathbf{x}$ is invertible. Also there is a non-invertible element. Namely, applying Lemma 13.5 to the varieties $\mathbf{V}(\mathbf{S})$ and $\mathbf{V}(\mathbf{M})$, we obtain a term t such that $t^{\mathbf{S}}(x,y,z)$ depends only on y while $t^{\mathbf{M}}$ is Maltsev. Let $s(x) = t(x,x,x)$. Now if $s^{\mathbf{U}}(\mathbf{x})$ were invertible, say $\mathbf{V}(\mathbf{S}) \models r(s(x)) \approx x$, then choosing $b \in B$ we would have a 0-projection $r^{\mathbf{B}}(t^{\mathbf{B}}(x,x,b))$ of $\mathbf{B}$. We conclude that the algebra $\mathbf{U}/\sim$ has precisely two elements—$\mathbf{x}/\sim$ and $s^{\mathbf{U}}(\mathbf{x})/\sim$. We shall complete the proof of this theorem by demonstrating the following claim.

Claim. The pair $\langle \mathcal{S}, \mathcal{M} \rangle$, where $\mathcal{S} = \mathbf{V}(\mathbf{U}/\sim)$ and $\mathcal{M} = \mathbf{V}(\mathbf{M})$, is strange.

We shall see that the four conditions in Definition 13.9 are consequences of the statements (1)-(5) above.

First, (1) of the definition holds because $\mathcal{S} \vee \mathcal{M}$ is a subvariety of $\mathbf{V}(\mathbf{B})$. Second, (2) of the definition follows from the statement (3) above. Third, from our construction above, it is easily seen that $\mathbf{F}_{\mathcal{S}}(\mathbf{x}) \cong \mathbf{U}/\sim$, and then statement (3) of the definition readily follows.

Finally, we prove condition (4) of Definition 13.9. Let $t(x,y,z)$ be the term provided by Lemma 13.5. As we noted above, this term is $\mathcal{S}$-constant, while $\mathcal{M} \models t(x,x,x) \approx x$. To prove the other assertion in Definition 13.9 (4), let $t(x,\bar{y})$ be any n-ary $\mathcal{S}$-normal term, so that $\mathcal{S} \models t(x,\bar{y}) \approx x$. Then $t^{\mathbf{S}}$ must depend on its first variable, but no other, since $\mathbf{S}$ is essentially unary. Moreover, $\mathbf{U}/\sim \models t(x,\bar{y}) \approx x$ means that $s^{\mathbf{U}}(\mathbf{x})$ is an invertible element, where $s(x) = t(x,x,\ldots,x)$. Letting $t^{(1)} = t$ and inductively, $t^{(k+1)}(x,\bar{y}) = t(t^{(k)}(x,\bar{y}),\bar{y})$, choose $m > 1$ such that where $\gamma(x,\bar{y}) = t^{(m)}(x,\bar{y})$ we have

$$\mathbf{B} \models \gamma(\gamma(x,\bar{y}),\bar{y}) \approx \gamma(x,\bar{y}).$$

Then

$$\mathbf{S} \models \gamma(x,\bar{y}) \approx s^m(x) \approx x$$

since $s^{\mathbf{S}}(x)$ is invertible. The two displayed formulas imply that for every $\bar{b} \in B^{n-1}$ the function $\gamma^{\mathbf{B}}(x,\bar{b})$ is 0-transparent and idempotent. Then statement (4) above implies that $\gamma^{\mathbf{B}}(x,\bar{b}) = x$ for all x; hence the equation $\gamma(x,\bar{y}) \approx x$ is an identity of $\mathbf{B}$, and therefore also an identity of $\mathbf{M}$. So if we define $r(x,\bar{y}) = t^{(m-1)}(x,\bar{y})$, then

$$\mathbf{M} \models r(t(x,\bar{y}),\bar{y}) \approx \gamma(x,\bar{y}) \approx x.$$

This completes our demonstration that $\langle \mathcal{S}, \mathcal{M} \rangle$ is strange, and completes our proof of the theorem. □

Our final task is to prove that strange pairs of varieties do not exist. Suppose that we do have a strange pair $\langle \mathcal{S}, \mathcal{M} \rangle$. Then $\mathcal{M}$ is structured. Thus it follows from Part I that $\mathcal{M}$ is the join of a discriminator variety and an Abelian Maltsev variety—i.e., an affine variety. We shall prove, separately, that the affine and discriminator subvarieties of $\mathcal{M}$ are trivial, thus implying that $\mathcal{M}$ is trivial and contradicting condition (2) in the definition of a strange pair.

LEMMA 13.11 *If $\langle \mathcal{S}, \mathcal{M} \rangle$ is a strange pair, then $\mathcal{M}$ is not a discriminator variety.*

PROOF. We assume that $\mathcal{S}$ is a locally finite strongly Abelian variety and $\mathcal{M}$ is a discriminator variety, and the pair $\langle \mathcal{S}, \mathcal{M} \rangle$ satisfies conditions (2)-(4) of Definition 13.9. We shall prove that $\mathcal{S} \vee \mathcal{M}$ is unstructured, so that $\langle \mathcal{S}, \mathcal{M} \rangle$ is not strange. We shall do this by interpreting into $\mathcal{S} \vee \mathcal{M}$ the class of graphs having at least three vertices.

We choose a finite simple algebra $\mathbf{M} \in \mathcal{M}$ (which we can do because $\mathcal{M}$ is non-trivial). Letting $\mathbf{S}$ be the free algebra $\mathbf{F}_{\mathcal{S}}(1)$ of $\mathcal{S}$ and taking $\mathbf{Q} = \mathbf{S} \times \mathbf{M}$, we shall use algebras in $SP(\mathbf{Q})$ for the interpretation.

By the third condition of Definition 13.9, we can write $S = \{\mu, \nu\}$, where $\mu \neq \nu$ and for every $\mathcal{S}$-constant term t, the constant value of t in $\mathbf{S}$ is μ. From conditions (3) and (4) combined we have

$$\begin{array}{l}\text{For every term } t(x_0, \ldots, x_{n-1}), \text{ if } \nu \text{ belongs to the} \\ \text{range of } t^{\mathbf{S}} \text{ then there exists } i < n \text{ such that} \\ \mathcal{S} \models t(x_0, \ldots, x_{n-1}) \approx x_i \text{ and for every} \\ m_0, \ldots, m_{n-1} \in M, \text{ the polynomial function } f(x) \\ \text{obtained from } t^{\mathbf{M}} \text{ by replacing } x_j \text{ with } m_j \\ \text{for all } j \neq i \text{ while replacing } x_i \text{ with } x, \\ \text{is a permutation of } M.\end{array} \tag{13.2}$$

Let $r(x)$ be an $\mathcal{S}$-constant term with $\mathcal{M} \models r(x) \approx x$ (condition (4)) and let $e(x)$ denote the polynomial function $r^{\mathbf{Q}}(x)$. Thus

$$\text{for } \langle s, m \rangle \in Q \text{ we have } e(\langle s, m \rangle) = \langle \mu, m \rangle. \tag{13.3}$$

We choose a fixed pair $m_0 \neq m_1$ of distinct elements of M, and define four elements of Q:

$$0 = \langle \mu, m_0 \rangle\,,\ 1 = \langle \mu, m_1 \rangle\,,\ 0' = \langle \nu, m_0 \rangle\,,\ 1' = \langle \nu, m_1 \rangle.$$

Let $t(x, y, z)$ be a term such that $t^{\mathbf{M}}$ is the discriminator operation on M—i.e., such that $t^{\mathbf{M}}(x, y, z) = z$ if $x = y$, and $= x$ if $x \neq y$. (See Theorem 0.22.) We define several polynomial operations of $\mathbf{Q}$:

$$b(x) = e(t^{\mathbf{Q}}(0, t^{\mathbf{Q}}(0, x, 1), 1));$$

$$x \vee y = e(t^{\mathbf{Q}}(b(x), 0, b(y)));$$

$$x \wedge y = e(t^{\mathbf{Q}}(b(x), 1, b(y))).$$

Observe that

> Each of the operations $b, e, \vee, \wedge$ is of the form $s^{\mathbf{Q}}(x, 0, 1)$ or $s^{\mathbf{Q}}(x, y, 0, 1)$ for some term s—i.e., is definable as a polynomial using the constants $0, 1$.
>
> We have $b \in \mathrm{E}(\mathbf{Q})$ and $b(Q)$ is just the set $\{0, 1\}$. The set $b(Q)$ is closed under $\vee$ and $\wedge$ and , with these operations, is a two-element lattice. (13.4)
>
> Finally, an element $x = \langle s, m\rangle \in Q$ satisfies $b(x) = e(x)$ iff $m \in \{m_0, m_1\}$.

Let $\mathbf{G} = \langle G, E\rangle$ be any graph with $|G| \geq 3$. We assume that G and E are disjoint sets. We define

$$T = \{a_0, a_1\} \cup G \cup E$$

where $\{a_0, a_1\}$ is some two-element set disjoint from $G \cup E$. The algebra into which we shall interpret $\mathbf{G}$ will be a subalgebra of $\mathbf{Q}^T$ generated by certain functions which we now define. First, let χ be the function in $\{0, 1\}^T$ such that $\chi^{-1}(0) = \{a_0\}$. Then for $g \in G$ define $\chi_g \in Q^T$ so that for $z \in T$ we have

$$\chi_g(z) = \begin{cases} 0' & z = a_0 \\ 1 & z = a_1 \\ 0 & z \in G - \{g\} \\ 1 & z = g \\ 0 & z \in E, g \notin z \\ 1 & z \in E, g \in z. \end{cases}$$

Then define

> $\mathbf{D}$ is the subalgebra of $\mathbf{Q}^T$ generated by the two constant functions $\hat{0}$ and $\hat{1}$ and the functions χ and $\chi_g (g \in G)$ defined above. (13.5)

The four operations on D that act coordinatewise like e, b, $\vee$, and $\wedge$ are, by (13.4), polynomial operations of $\mathbf{D}$. We shall denote each of these operations by the same symbol used for the corresponding operation in Q. These operations can be defined in $\mathbf{D}$ by first order formulas using the elements $\hat{0}$ and $\hat{1}$ as parameters. Thus the sets

$$\Delta = b(D)\,,\ \Lambda = \{\alpha \in D : b(\alpha) = e(\alpha)\}$$

are definable in $\mathbf{D}$. It is obvious from (13.3) and (13.4) that

$$\Delta = D(\{0,1\}) = D \cap \{0,1\}^T \text{ and } \Delta \text{ is a sublattice of } \langle\{0,1\}, \vee, \wedge\rangle^T; \text{ moreover, } \Lambda = D \cap \{0,0',1,1'\}^T.$$

Notice that all of the generating functions of $\mathbf{D}$ belong to the definable set Λ. We define

$$\Lambda^* = \{\alpha \in \Lambda : \alpha \neq e(\alpha) \text{ and } e(\alpha) \leq \chi\},$$

where $e(\alpha) \leq \chi$ means, of course, that $e(\alpha) \vee \chi = \chi$.

The definable set Λ^* contains the elements χ_g ($g \in G$). The next claim is the crux of this proof.

Claim. If $\alpha \in \Lambda^*$ then $e(\alpha) = e(\chi_g)$ for some $g \in G$.

To prove this claim, suppose that $\alpha \in \Lambda^*$. We can choose (by (13.5)) a term $F(x_0, \ldots, x_{n-1}, w, u, v)$ and elements $g_0, \ldots, g_{n-1} \in G$ such that where $\chi_i = \chi_{g_i}$ for $i < n$, we have

$$\alpha = F^{\mathbf{D}}(\chi_0, \ldots, \chi_{n-1}, \chi, \hat{0}, \hat{1}).$$

If $z \in T - \{a_0\}$ then $p_0(\alpha(z)) = \mu$—since $p_0(\tau(z)) = \mu$ for $\tau = \chi_i, \chi, \hat{0}, \hat{1}$ and $F^{\mathbf{S}}(\mu, \ldots, \mu) = \mu$ by (13.2). Thus since $\alpha \neq e(\alpha)$, implying $p_0(\alpha(z)) = \nu$ for some z, we must have that $p_0(\alpha(a_0)) = \nu$. This means that

$$F^{\mathbf{S}}(\nu, \ldots, \nu, \mu, \mu, \mu) = \nu,$$

which implies by (13.2) that for some $i < n$ we have

$$\mathcal{S} \models F(x_0, \ldots, x_{n-1}, w, u, v) \approx x_i.$$

Now by changing our ordering of $\chi_0, \ldots, \chi_{n-1}$ if necessary, and permuting the variables of F to match, we can arrange that $i = 0$, so that F is $\mathcal{S}$-normal. We shall now show that $e(\alpha) = e(\chi_0)$, completing the proof of the claim.

By $\bar{k}(z)$, where $z \in T$, we denote the $n+3$-tuple

$$\langle p_1(\chi_0(z)), \ldots, p_1(\chi_{n-1}(z)), p_1(\chi(z)), m_0, m_1\rangle.$$

For $z \in T$ we have

$$p_1(\alpha(z)) = F^{\mathbf{M}}(\bar{k}(z)) \in \{m_0, m_1\}$$

since $\alpha \in \Lambda$. Now $e(\alpha) \leq \chi$ means that $e(\alpha)(a_0) = 0$, i.e., that $F^{\mathbf{M}}(\bar{k}(a_0)) = m_0$. Notice that

$$\bar{k}(a_0) = \langle m_0, \ldots, m_0, m_0, m_0, m_1\rangle, \text{ also}$$

$$\bar{k}(a_1) = \langle m_1, \ldots, m_1, m_1, m_0, m_1\rangle.$$

Since F is $\mathcal{S}$-normal, the term $H(x,y,u) = F(x,\ldots,x,y,u)$ is $\mathcal{S}$-normal as well. Thus by (13.2) we have $F^{\mathbf{M}}(\bar{k}(a_1)) \neq F^{\mathbf{M}}(\bar{k}(a_0))$, implying that $F^{\mathbf{M}}(\bar{k}(a_1)) = p_1(\alpha(a_1)) = m_1$. Now in order to prove that $e(\alpha) = e(\chi_0)$, suppose that $z \in T$ and $e(\alpha)(z) \neq e(\chi_0)(z)$. This means that $F^{\mathbf{M}}(\bar{k}(z)) \neq p_1(\chi_0(z))$. Consider first the case that $p_1(\chi_0(z)) = m_0$ and $F^{\mathbf{M}}(\bar{k}(z)) = m_1$. We let $J(x,y,u,v)$ be the term that results upon replacing in the term $F(x_0,\ldots,x_{n-1},w,u,v)$ each variable x_i such that $p_1(\chi_i(z)) = m_0$ by x, and replacing w and every other variable x_i by y, while keeping u and v unchanged. Thus $\mathcal{S} \models J(x,y,u,v) \approx F(x,\ldots,x)$ so J is $\mathcal{S}$-normal. But we have

$$J^{\mathbf{M}}(m_0,m_1,m_0,m_1) = F^{\mathbf{M}}(\bar{k}(z)) = m_1, \text{ while}$$

$$J^{\mathbf{M}}(m_1,m_1,m_0,m_1) = F^{\mathbf{M}}(\bar{k}(a_1)) = m_1$$

which contradicts (13.2). Thus we have a contradiction in this case. The case where $p_1(\chi_0(z)) = m_1$ and $F^{\mathbf{M}}(\bar{k}(z)) = m_0 = F^{\mathbf{M}}(\bar{k}(a_0))$ can be handled in a similar manner, taking $J(x,y,u,v)$ to be the term that results upon replacing in $F(x_0,\ldots,x_{n-1},w,u,v)$ the variable w and every variable x_i such that $p_1(\chi_i(z)) = m_1$ by x, and all other variables x_i by y. Thus we conclude our proof of the claim.

To finish the proof, we take $G^\star = \{e(\alpha) : \alpha \in \Lambda^\star\}$. By the claim, $G^\star = \{e(\chi_g) : g \in G\}$ and $G^\star$ is a definable subset in $\mathbf{D}$. We define, for $g \in G$, $\varphi(g) = e(\chi_g)$, so that φ is a bijection of G with $G^\star$. It is easy to see that when $g_1, g_2 \in G$ we have $\{g_1,g_2\} \in E$ iff $\varphi(g_1) \neq \varphi(g_2)$ and $\varphi(g_1) \wedge \varphi(g_2) \not\leq \varphi(h)$ for any $h \in G - \{g_1,g_2\}$. (This is why we required $\mathbf{G}$ to have more than two vertices.) In fact, if $g_1 \neq g_2$ and $\{g_1,g_2\} \notin E$ then $\varphi(g_1) \wedge \varphi(g_2)$ is the function in $\{0,1\}^T$ which is 1 only at $z = a_1$; whereas if $\{g_1,g_2\} \in E$ then $\varphi(g_1) \wedge \varphi(g_2)$ is 1 precisely at $z = a_1$ and $z = \{g_1,g_2\}$. Thus φ is an isomorphism of $\langle G, E\rangle$ with a structure $\langle G^\star, E^\star\rangle$, which has been shown to be definable in $\mathbf{D}$. Our proof is complete. □

LEMMA 13.12 *If $\langle \mathcal{S}, \mathcal{M}\rangle$ is a strange pair, then $\mathcal{M}$ is not an affine variety.*

This lemma will complete the work of this chapter. For the proof, we now assume that $\mathcal{S}$ is a locally finite strongly Abelian variety, $\mathcal{M}$ is a locally finite affine variety, and the pair $\langle \mathcal{S}, \mathcal{M}\rangle$ satisfies conditions (2)-(4) of Definition 13.9. We shall interpret the class of graphs into $\mathcal{S} \vee \mathcal{M}$, thus showing that Definition 13.9 (1) fails and $\langle \mathcal{S}, \mathcal{M}\rangle$ cannot be strange. This will require some preparation. Let $\mathbf{S}$ denote the free algebra on one generator in $\mathcal{S}$. By Definition 13.9 (3)-(4), we can choose an $\mathcal{S}$-constant term μ so that $\mathcal{M} \models \mu(x) \approx x$ and $S = \{\mu^{\mathbf{S}}(\mathbf{x}), \mathbf{x}\}$, where all $\mathcal{S}$-constant terms have only the value $\mu^{\mathbf{S}}(\mathbf{x})$ in $\mathbf{S}$.

Let $\mathbf{F}_2$ denote the free algebra in $\mathcal{M}$ freely generated by $\mathbf{x}$ and $\mathbf{y}$. If t is any term, then $\mathcal{M} \models t \approx \mu(t)$ while $\mu(t)$ is $\mathcal{S}$-constant. Thus we can choose a Maltsev term d for $\mathcal{M}$ that is $\mathcal{S}$-constant. Recall from Theorem 0.19 that algebras in $\mathcal{M}$ are polynomially equivalent to unitary modules over a certain ring $\mathbf{R}$. Using the term d and the algebra $\mathbf{F}_2$, this ring is constructed as follows. [See Burris, McKenzie, Chapter 10; or Freese, McKenzie, Theorems 9.12–9.19.]

$$R = \{r^{\mathbf{F}_2}(\mathbf{x}, \mathbf{y}) \in F_2 : \mathcal{M} \models r(x, x) \approx x\} ;$$

$$0_{\mathbf{R}} = \mathbf{y} ,\ 1_{\mathbf{R}} = \mathbf{x} ;$$

$$r^{\mathbf{F}_2}(\mathbf{x}, \mathbf{y}) + s^{\mathbf{F}_2}(\mathbf{x}, \mathbf{y}) = d^{\mathbf{F}_2}(r^{\mathbf{F}_2}(\mathbf{x}, \mathbf{y}), \mathbf{y}, s^{\mathbf{F}_2}(\mathbf{x}, \mathbf{y})) ;$$

$$r^{\mathbf{F}_2}(\mathbf{x}, \mathbf{y}) \cdot s^{\mathbf{F}_2}(\mathbf{x}, \mathbf{y}) = r^{\mathbf{F}_2}(s^{\mathbf{F}_2}(\mathbf{x}, \mathbf{y}), \mathbf{y}) ;$$

$$-r^{\mathbf{F}_2}(\mathbf{x}, \mathbf{y}) = d^{\mathbf{F}_2}(\mathbf{y}, r^{\mathbf{F}_2}(\mathbf{x}, \mathbf{y}), \mathbf{y}) .$$

Strictly speaking, the elements of R are members of $\mathbf{F}_2$, but often we allow ourselves to regard them as binary terms in the language of $\mathcal{M}$.

$\mathbf{F}_2$ becomes an $\mathbf{R}$-module upon defining addition and subtraction to be

$$a + b = d^{\mathbf{F}_2}(a, \mathbf{y}, b) ;$$

$$-a = d^{\mathbf{F}_2}(\mathbf{y}, a, \mathbf{y});$$

$$0 = \mathbf{y} ;$$

and for $r \in R$,

$$r \cdot a = r^{\mathbf{F}_2}(a, \mathbf{y}).$$

The structure so defined is an $\mathbf{R}$-module, which we denote by $\mathbf{M}(\mathbf{F}_2, \mathbf{y})$. Observe that for r, $s \in R$ we have $r = s$ iff $r \cdot \mathbf{x} = s \cdot \mathbf{x}$ in this module. (We shall need this observation later.)

The module $\mathbf{M}(\mathbf{F}_2, \mathbf{y})$ is polynomially equivalent to $\mathbf{F}_2$. In fact, the operation $d^{\mathbf{F}_2}(x, y, z)$ is equal to the operation $x - y + z$ expressed in module operations; and for any term $t(x_0, \ldots, x_{n-1})$ in the language of $\mathbf{M}$ there are unique elements $r_0, \ldots, r_{n-1} \in R$ such that

$$\mathbf{F}_2 \models t(x_0, \ldots, x_{n-1}) \approx r_0 \cdot x_0 + \cdots + r_{n-1} \cdot x_{n-1} + t_\delta(0)$$

where $0 = \mathbf{y}$ and $t_\delta(z) = t(z, \ldots, z)$.

By the $\mathcal{M}$-**representation** of the term t we mean the expression

$$r_0 \cdot x_0 + \cdots + r_{n-1} \cdot x_{n-1} + t_\delta(0).$$

Recall that t is $\mathcal{S}$-**normal** iff $\mathcal{S} \models t(x_0, \ldots, x_l) \approx x_0$ (confer Definition 13.9). Let $\kappa = |R|$.

Definition 13.13 Let $t(x_0, x_1, \ldots, x_l)$ be a term with $\mathcal{M}$-representation $r_0 \cdot x_0 + \cdots + r_l \cdot x_l + t_\delta(0)$.

(1) We call $t(x_0, \ldots, x_l)$ **reduced** if $l \leq \kappa + 1$ and in $\mathcal{S}$, t depends at most on its first variable x_0.

(2) For $\mathcal{S}$-normal t we denote r_0 by r_t, and put S_t equal to the right $\mathbf{R}$ ideal generated by $\{r_i : i > 0\}$.

(3) Let $p(x_0, \ldots, x_m)$ and $q(x_0, \ldots, x_k)$ be two terms. The term

$$p(q(x_0, \ldots, x_k), x_{k+1}, \ldots, x_{k+m})$$

will be denoted by $p \circ q$.

LEMMA 13.14

(i) *$p \circ q$ is an $\mathcal{S}$-normal term iff p and q are both $\mathcal{S}$-normal terms. If these terms are $\mathcal{S}$-normal, then $r_p \cdot r_q = r_{p \circ q}$ and $S_{p \circ q} = r_p \cdot S_q + S_p$.*

(ii) *If $t = t(x_0, x_1, \ldots, x_k)$ is an $\mathcal{S}$-normal term, then there exists an $\mathcal{S}$-normal term $s = s(x_0, x_1, \ldots, x_k)$ such that*

$$\mathcal{M} \models s(t(x_0, \ldots, x_k), x_1, \ldots, x_k) \approx x_0,$$

$$\mathcal{M} \models t(s(x_0, \ldots, x_k), x_1, \ldots, x_k) \approx x_0,$$

$r_s \cdot r_t = r_t \cdot r_s = 1$, and $r_s \cdot S_t = S_s$. Thus r_t is invertible.

(iii) *There exists an $\mathcal{S}$-normal term t such that $r_t = 1$ and $S_p \subseteq S_t \neq R$ for every $\mathcal{S}$-normal term p. Thus if p and q are $\mathcal{S}$-normal terms then $S_p + S_q \neq R$.*

PROOF. Statement (i) is obvious from the definition of an $\mathcal{S}$-normal term and from Definition 13.13. To prove (ii) let t be $\mathcal{S}$-normal and let s be the $\mathcal{S}$-normal term supplied by Definition 13.9 (4), so that

$$\mathcal{M} \models s(t(x_0, \ldots, x_k), x_1, \ldots, x_k) \approx x_0.$$

Since $\mathcal{M}$ is locally finite, the same equation with s and t interchanged is valid in $\mathcal{M}$. Letting $r_t x_0 + t_1 x_1 + \cdots + t_k x_k + t_\delta(0)$ and $r_s x_0 + s_1 x_1 + \cdots + s_k x_k + s_\delta(0)$ be the $\mathcal{M}$-representations of t and s, we calculate that

$$r_s r_t \cdot x_0 + u_1 x_1 + \cdots + u_k x_k + r_s t_\delta(0) + s_\delta(0)$$

is an $\mathcal{M}$-representation of the term x_0, where $u_i = r_s t_i + s_i$. Since representations are unique, we have that $r_s r_t = 1$ and $r_s t_i = -s_i$ $(1 \leq i \leq k)$. Statement (ii) follows from these considerations.

To prove (iii), we use the fact that $\mathbf{R}$ is a finite ring. Let $t = t(x_0, \ldots, x_k)$ be an $\mathcal{S}$-normal term, with $\mathcal{M}$-representation $r_t x_0 + t_1 x_1 + \cdots + t_\delta(0)$, such that S_t is maximal for $\mathcal{S}$-normal terms. Letting s be the term supplied by statement (ii), note that $S_{t \circ s} = r_t S_s + S_t$ by (i), and $r_{t \circ s} = 1$. Thus we can assume that $r_t = 1$. Now if p is any $\mathcal{S}$-normal term then $t \circ p$ is $\mathcal{S}$-normal and $S_{t \circ p} = r_t S_p + S_t = S_p + S_t$; thus $S_p \subseteq S_t$ since S_t is maximal. Finally, we must show that $S_t \neq R$. Suppose to the contrary that there exist $t_1', \ldots, t_k' \in R$ such that $1 = t_1 t_1' + \cdots + t_k t_k'$. Letting z be a new variable, we note that each term $t_i'(x, z)$ has the representation $t_i' x + (1 - t_i')z$, while $d(x, y, z)$ has the representation $x - y + z$. Thus $u_i(x, z) = d(z, t_i'(x, z), z)$ has the representation $-t_i' x + (1 + t_i')z$. We define a term

$$s(x, z) = t(x, u_1(x, z), \ldots, u_k(x, z)).$$

It is easily verified that $s(x, z)$ has the representation

$$(1 - \sum t_i t_i')x + \sum t_i(1 + t_i')z + s_\delta(0);$$

i.e., $r_s = 0$. But s is clearly $\mathcal{S}$-normal. So we have a contradiction to the fact proved above, that r_s is a unit in $\mathbf{R}$ whenever s is $\mathcal{S}$-normal. □

Proof of Lemma 13.12. We shall interpret the class of graphs into the class of subalgebras of powers of $\mathbf{S} \times \mathbf{F}_2$. Let $\mathbf{Q} = \mathbf{S} \times \mathbf{F}_2$ and let $\mathbf{G} = \langle G, E \rangle$ be a graph. Choose an element a not in G, and let $T = \{a\} \cup G$. Recall that $S = \{\mathbf{x}, \mu^{\mathbf{S}}(\mathbf{x})\}$ and that $\mathbf{F}_2$ is freely generated by $\mathbf{x}$ and $\mathbf{y}$. Also, $\mathbf{y}$ is the zero element of the module $\mathbf{M}(\mathbf{F}_2, \mathbf{y})$.

We put $0 = \langle \mu^{\mathbf{S}}(\mathbf{x}), \mathbf{y} \rangle \in Q$, $1 = \langle \mathbf{x}, \mathbf{x} \rangle \in Q$, and define several functions belonging to Q^T. For $g \in G$ and $z \in T$ let

$$\chi_g(z) = \begin{cases} 1 & \text{if } z = g \\ 0 & \text{otherwise.} \end{cases}$$

For $e \in E$ and $z \in T$ let

$$\chi_e^1(z) = \begin{cases} 1 & \text{if } z \in e \\ 0 & \text{otherwise,} \end{cases}$$

and

$$\chi_e^2(z) = \begin{cases} 1 & \text{if } z \in e \text{ or } z = a \\ 0 & \text{otherwise.} \end{cases}$$

Let $\chi : T \to Q$ be the function

$$\chi(z) = \begin{cases} 1 & \text{if } z = a \\ 0 & \text{otherwise.} \end{cases}$$

Finally, let $G^\star = \{\chi_g : g \in G\}$, let $E^\star = \{\chi_e^i : e \in E \text{ and } i = 1, 2\}$, and let

$$\mathbf{D} = \mathrm{Sg}^{\mathbf{Q}^T}(G^\star \cup E^\star \cup \{\chi, \hat{0}\})$$

where $\hat{0}$ is the constant function in Q^T corresponding to $0 = \langle \mu^{\mathbf{S}}(\mathbf{x}), \mathbf{y} \rangle \in Q$. We use p_0 and p_1 as usual to denote the projections of $\mathbf{Q}$ onto $\mathbf{S}$ and $\mathbf{F}_2$ respectively; thus $p_i f$ (where $f \in D$) belongs to $\mathbf{S}^T$ or $\mathbf{F}_2^T$.

Now since $\mathcal{S} \vee \mathcal{M}$ is locally finite, there are, modulo equality in $\mathcal{S} \vee \mathcal{M}$, only a finite number of reduced terms (confer Definition 13.13 (1)). Let T_r be a finite set of reduced terms such that every reduced term is equal in $\mathcal{S} \vee \mathcal{M}$ to a member of T_r; and let $\mathrm{T}_{rn} \subseteq \mathrm{T}_r$ be the set of $\mathcal{S}$-normal terms belonging to T_r. Notice that the definable set $\mu(D)$ (where μ is the $\mathcal{S}$-constant term used above) is just the set of all $f \in D$ such that $p_0 f(z) = \mu^{\mathbf{S}}(\mathbf{x})$ for all $z \in T$, and that $\mu(D)$ is a subalgebra of $\mathbf{D}$ isomorphic to the projection of $\mathbf{D}$ into $\mathbf{F}_2^T$. Since $\mathbf{F}_2^T$ is polynomially equivalent to a module over $\mathbf{R}$, then $\mu(D)$ has the structure of a module over $\mathbf{R}$ in which addition is the restriction to $\mu(D)$ of the polynomial operation $u + v = d^{\mathbf{D}}(u, \hat{0}, v)$ of $\mathbf{D}$.

Claim 1. Let $t(x_0, \ldots, x_n)$ be a term such that in $\mathcal{S}$ it depends on at most the variable x_0 and let $f_i \in D$ for $i \leq n$. Then there exists a reduced term $s(x_0, \ldots x_l) \in \mathrm{T}_r$ and elements $g_j \in D$ for $j > 0$ such that

$$t^{\mathbf{D}}(f_0, \ldots, f_n) = s^{\mathbf{D}}(f_0, g_1, \ldots, g_l),$$

$r_t = r_s$, $S_t = S_s$, and $s \in \mathrm{T}_{rn}$ iff t is $\mathcal{S}$-normal.

To prove this claim, let $r_t x_0 + \sum t_i x_i + t_\delta(\mathbf{y})$ be the $\mathcal{M}$-representation of t. If $n \leq \kappa$ $(= |R|)$ then t is reduced and we can just let s be a member of T_r equivalent to t in $\mathcal{S} \vee \mathcal{M}$. Otherwise, we can find $i \neq j$ such that $t_i = t_j$. Let $g_i = f_i + f_j$ $(= d^{\mathbf{D}}(f_i, \hat{0}, f_j))$ and let $g_j = \hat{0}$. Then it follows that by replacing f_i by g_i and f_j by g_j in t, we get the same value in $\mathbf{D}$. By repeating this process (and perhaps rearranging some of the variables of t), we can find an $m \leq \kappa$ and $g_j \in D$ for $j \leq m$ such that

$$t^{\mathbf{D}}(f_0, \ldots, f_n) = t^{\mathbf{D}}(f_0, g_1, \ldots, g_m, \hat{0}, \hat{0}, \ldots, \hat{0}).$$

Thus we can choose our term s to be a specialization of t.

Now we give a first order definition for what will turn out to be an equivalence relation on $D - \mu(D)$. For f, $g \in D - \mu(D)$ we put

$$f \sim g \text{ iff } g = t^{\mathbf{D}}(f, h_1, \ldots, h_l) \text{ for some } t \in \mathrm{T}_{rn} \text{ and}$$
$$h_1, \ldots, h_l \in D.$$

Claim 2. Let f, $g \in D - \mu(D)$. Then $f \sim g$ iff for some $h \in G^\star \cup E^\star \cup \{\chi\}$, $h \sim f$ and $h \sim g$. We have that $\sim$ is an equivalence relation on $D - \mu(D)$. Each equivalence class contains exactly one member of the set $G^\star \cup E^\star \cup \{\chi\}$.

To prove this claim, we note first that if $f \sim g$ then $p_0 f = p_0 g$, since if t is $\mathcal{S}$-normal then $\mathcal{S} \models t(x_0, \ldots, x_l) \approx x_0$. This establishes, among other things, that $f \sim g$ cannot hold if f and g are distinct members of the generating set of $\mathbf{D}$. Next, note that every element $f \in D$ can be expressed as $f = t^{\mathbf{D}}(h_0, \ldots, h_n)$ for some term t where $h_0, \ldots, h_n$ are members of the generating set for $\mathbf{D}$. Assume that $f \notin \mu(D)$. Then t is not $\mathcal{S}$-constant, and in fact t must depend, in $\mathcal{S}$, on a variable x_i such that $h_i \neq \hat{0}$. Thus, permuting the variables, we have $f = s^{\mathbf{D}}(h, g_1, \ldots, g_n)$ for an $\mathcal{S}$-normal term s, some elements $g_1, \ldots, g_n \in D$, and an element $h \in G^\star \cup E^\star \cup \{\chi\}$. By Claim 1, we can assume that $s \in \mathrm{T}_{rn}$; hence $h \sim f$. Now if $f \sim g$ then $p_0 g = p_0 f = p_0 h$, hence the $h' \in G^\star \cup E^\star \cup \{\chi\}$ with $h' \sim g$ must equal h. Thus $f \sim g$ implies that $h \sim f$ and $h \sim g$ for some $h \in G^\star \cup E^\star \cup \{\chi\}$.

Now suppose that f, g, $h \in D - \mu(D)$ and $h \sim f$ and $h \sim g$. We shall show that $f \sim g$. The symmetry and transitivity of $\sim$ will then follow readily, and in fact all parts of the claim should then become obvious. Assume that $f = t^{\mathbf{D}}(h, g_1, \ldots, g_l)$ and $g = s^{\mathbf{D}}(h, h_1, \ldots, h_k)$ where s, $t \in \mathrm{T}_{rn}$. By Lemma 13.14 (ii) there is an $\mathcal{S}$-normal r such that $h = r^{\mathbf{D}}(f, g_1, \ldots, g_l)$. Then $g = s_\circ r^{\mathbf{D}}(f, g_1, \ldots, g_l, h_1, \ldots, h_k)$; and by Claim 1 we find that $f \sim g$.

In order to see that the set $G^\star/\sim$ is definable, we shall have to define $E^\star/\sim$. In the formulas we now construct, the element χ is used as a parameter. Let $\mathrm{Edg}(u, v)$ be the formula

$$\neg u \sim v \wedge (\exists u' \sim u)(\exists v' \sim v)$$
$$(u' + \chi \approx \mu(v') \vee v' + \chi \approx \mu(u')).$$

Claim 3. $\mathbf{D} \models \mathrm{Edg}(f, g)$ iff there exists $e \in E$ such that $f \sim \chi_e^i$ and $g \sim \chi_e^j$ for some i, j such that $\{i, j\} = \{1, 2\}$.

To prove this claim, we note first that $\chi_e^1 + \chi = \mu^{\mathbf{D}}(\chi_e^2)$, so $\mathbf{D} \models \mathrm{Edg}(\chi_e^1, \chi_e^2)$ and $\mathbf{D} \models \mathrm{Edg}(\chi_e^2, \chi_e^1)$. Now suppose that $\mathbf{D} \models \mathrm{Edg}(f, g)$. Then choose $f', g' \in D$ and $f^\circ, g^\circ \in G^\star \cup E^\star \cup \{\chi\}$ such that $f \sim f' \sim f^\circ$ and $g \sim g' \sim g^\circ$ and $f' + \chi = \mu^{\mathbf{D}}(g')$ or $g' + \chi = \mu^{\mathbf{D}}(f')$. We can assume that $f' + \chi = \mu^{\mathbf{D}}(g')$. Note that $f^\circ \neq g^\circ$ since $\neg f \sim g$. We next choose s, $t \in \mathrm{T}_{rn}$ such that $f' = s^{\mathbf{D}}(f^\circ, f_1, \ldots, f_l)$ and $g' = t^{\mathbf{D}}(g^\circ, g_1, \ldots, g_k)$ (f_i, $g_j \in D$). Let $r_s x_0 + s_1 x_1 + \cdots + s_l x_l + s_\delta(\mathbf{y})$ and $r_t x_0 + t_1 x_1 + \cdots + t_k x_k + t_\delta(\mathbf{y})$ be the respective $\mathcal{M}$-representations of s and t. Our objective is to prove that $f^\circ(z) = g^\circ(z)$ for all $z \in T - \{a\}$. This will finish the proof of this claim, since the only pairs of distinct members of $G^\star \cup E^\star \cup \{\chi\}$ that agree on $T - \{a\}$ are those pairs of the form χ_e^1, χ_e^2.

So let $z \in T$, $z \neq a$, and suppose that $f^\circ(z) \neq g^\circ(z)$, say $f^\circ(z) = \langle \mathbf{x}, \mathbf{x} \rangle = 1$ and $g^\circ(z) = \langle \mu^{\mathbf{S}}(\mathbf{x}), \mathbf{y} \rangle = 0$. (The other case will be seen to lead to the same contradiction.) Since $\chi(z) = 0$, we have that

$$p_1 f'(z) = p_1(f' + \chi)(z) = p_1 g'(z).$$

This means that

$$r_s \mathbf{x} + s_1 p_1 f_1(z) + \cdots + s_l \cdot p_1 f_l(z) + s_\delta^{\mathbf{F}_2}(\mathbf{y}) =$$

$$r_t \mathbf{y} + t_1 p_1 g_1(z) + \cdots + t_k \cdot p_1 g_k(z) + t_\delta^{\mathbf{F}_2}(\mathbf{y}).$$

Note that every element of F_2 can be expressed in the form $r\mathbf{x} + p^{\mathbf{F}_2}(\mathbf{y})$ for some $r \in R$ and some unary term p. Expressing the elements $p_1 f_1(z), \ldots, p_1 g_k(z)$ in this way, we can rewrite the above equation in the module $\mathbf{M}(\mathbf{F}_2, \mathbf{y})$ as

$$r_s \mathbf{x} = (\gamma + \lambda)\mathbf{x} + p^{\mathbf{F}_2}(\mathbf{y})$$

with p a unary term, and $\gamma \in S_s$ and $\lambda \in S_t$. Now the endomorphism of $\mathbf{F}_2$ that sends $\mathbf{x}$ to $\mathbf{y}$ and $\mathbf{y}$ to $\mathbf{y}$ preserves module operations, and so substituting $\mathbf{y}$ for $\mathbf{x}$ in the displayed equation leads to

$$p^{\mathbf{F}_2}(\mathbf{y}) = \mathbf{y}$$

$$(r_s - \gamma - \lambda)\mathbf{x} = 0 \ (= \mathbf{y}).$$

This implies that $r_s = \gamma + \lambda \in S_s + S_t$; but that contradicts Lemma 13.14 (ii)-(iii).

Let $\mathrm{Un}(z)$ be the formula

$$\mu(z) \not\approx z \wedge \neg(z \sim \chi) \wedge \neg(\exists v)(\mathrm{Edg}(z, v)).$$

It follows by Claims 2 and 3 that $\mathbf{D} \models \mathrm{Un}(f)$ iff $f \sim \chi_g$ for some $g \in G$.

Finally, let $\mathrm{E}(u,v)$ be the formula

$$\mathrm{Un}(u) \wedge \mathrm{Un}(v) \wedge \neg(u \sim v) \wedge (\exists u', v', z, w)$$

$$(\mathrm{Edg}(z,w) \wedge u' \sim u \wedge v' \sim v \wedge u' + v' \approx \mu(z)).$$

Claim 4. $\mathbf{D} \models \mathrm{E}(\alpha,\beta)$ if and only if $\alpha \sim \chi_g$ and $\beta \sim \chi_h$ for some g, $h \in G$ such that $\{g,h\} \in E$.

To prove this claim in one direction, notice that if $\{g,h\} \in E$ and $\alpha \sim \chi_g$ and $\beta \sim \chi_h$, then we clearly can fulfill the clauses of $\mathrm{E}(\alpha,\beta)$ by taking $u' = \chi_g$, $v' = \chi_h$, $z = \chi_e^1$, and $w = \chi_e^2$. Conversely, suppose that $\mathbf{D} \models \mathrm{E}(\alpha,\beta)$; say $\alpha \sim \alpha'$, $\beta \sim \beta'$, $\neg\alpha \sim \beta$, and $\alpha' + \beta' = \mu^{\mathbf{D}}(\delta)$ where $\mathbf{D} \models \mathrm{Un}(\alpha) \wedge \mathrm{Un}(\beta) \wedge \mathrm{Edg}(\delta,\gamma)$. Thus by Claim 3, we have $\alpha \sim \alpha' \sim \chi_g$ and $\beta \sim \beta' \sim \chi_h$ where $g \neq h$, and $\delta \sim \chi_e^i$ and $\gamma \sim \chi_e^j$ for some $e \in E$ and $\{i,j\} = \{1,2\}$. We wish to prove that $e = \{g,h\}$. Thus we assume that this fails, and work toward a contradiction.

Since $e \neq \{g,h\}$ then we can choose $g' \in e - \{g,h\}$. We choose r, s, $t \in \mathrm{T}_{rn}$ and $\alpha_1, \ldots, \delta_m \in D$ such that

$$\begin{aligned} \alpha' &= r^{\mathbf{D}}(\chi_g, \alpha_1, \ldots, \alpha_k) \\ \beta' &= s^{\mathbf{D}}(\chi_h, \beta_1, \ldots, \beta_l) \\ \delta &= t^{\mathbf{D}}(\chi_e^i, \delta_1, \ldots, \delta_m). \end{aligned}$$

Let $a_j = p_1\alpha_j(g')$, $b_j = p_1\beta_j(g')$, $d_j = p_1\delta_j(g')$; and let

$$r_r x_0 + r_1 x_1 + \cdots$$

$$r_s x_0 + s_1 x_1 + \cdots$$

$$r_t x_0 + t_1 x_1 + \cdots$$

be the respective $\mathcal{M}$-representations of r, s, t. Using that $p_1\chi_g(g') = \mathbf{y} = p_1\chi_h(g')$ while $p_1\chi_e^i(g') = \mathbf{x}$ and $p_1\alpha' + p_1\beta' = p_1\delta$, we derive the equation

$$r_r\mathbf{y} + r_1a_1 + \cdots + r_ka_k + r_s\mathbf{y} + s_1b_1 + \cdots + s_lb_l + p^{\mathbf{F}_2}(\mathbf{y}) =$$

$$r_t\mathbf{x} + t_1d_1 + \cdots + t_md_m + q^{\mathbf{F}_2}(\mathbf{y}).$$

Just as in the proof of Claim 3, this leads to the conclusion that $r_t \in S_r + S_s + S_t$, and this contradicts Lemma 13.14 (iii). Thus the proof of Claim 4 is complete.

It should now be clear that the definable structure $\langle \mathrm{Un}^{\mathbf{D}}, \mathrm{E}^{\mathbf{D}} \rangle / \sim$ is isomorphic to the graph $\mathbf{G}$ via the map which sends $g \in G$ to the element $\chi_g/\sim$ in $\mathrm{Un}^{\mathbf{D}}/\sim$. □

THEOREM 13.15 *There does not exist a strange pair of varieties.*

PROOF. Suppose that $\langle \mathcal{S}, \mathcal{M} \rangle$ is a strange pair. Thus $\mathcal{M}$ is structured. Since $\mathcal{M}$ is Maltsev, it has no nontrivial strongly Abelian algebras. Thus it follows from Theorems 1.4, 4.1 and 5.4 (or Theorem 13.7) that $\mathcal{M} = \mathcal{M}_2 \vee \mathcal{M}_3$ where $\mathcal{M}_2$ is affine and $\mathcal{M}_3$ is a discriminator variety. Either $\mathcal{M}_2$ or $\mathcal{M}_3$ must be nontrivial. On the other hand, it is clear that if $\mathcal{M}_i$ ($i \in \{2,3\}$) is nontrivial, then $\langle \mathcal{S}, \mathcal{M}_i \rangle$ is a strange pair. Hence we have a contradiction either to Lemma 13.11 or to Lemma 13.12. □

Chapter 14

Conclusion

For any class of similar structures we have the implications

$$\text{unstructured} \Rightarrow \text{hereditarily undecidable} \Rightarrow \text{undecidable}.$$

According to Theorem 13.10, the family of unstructured varieties includes every locally finite variety that fails to decompose as the product of a strongly Abelian variety, an affine variety, and a discriminator variety. Consequently, a decidable locally finite variety must decompose in this fashion, and this reduces the study of decidable locally finite varieties to the examination of decidable locally finite varieties that fall into one of the three special cases. For locally finite, strongly Abelian varieties, we presented in Chapters 11 and 12 a simple necessary and sufficient criterion for decidability. We saw that the implications displayed above are reversible for varieties of this kind.

For locally finite discriminator varieties, no criterion for decidability is known. We begin our list of open problems with this one.

PROBLEM 1: *Which locally finite discriminator varieties are undecidable (unstructured, ω-unstructured)?*

S. Burris and H. Werner [1979] proved that every finitely generated discriminator variety of finite type is decidable. Some more recent results on Problem 1 can be found in S. Burris [1985] and S. Burris, R. McKenzie and M. Valeriote [1989].

The decidability question for locally finite affine varieties is very interesting and seems to be very difficult. Corresponding to an affine variety $\mathcal{V}$ there is a ring $\mathbf{R}$, which we examined in our proof of Lemma 13.12. According to Burris and McKenzie [1981] Theorem 10.6, for locally finite $\mathcal{V}$, the variety of left unitary modules over $\mathbf{R}$ is decidable if and only if $\mathcal{V}$ is decidable. Thus we have the next problem.

PROBLEM 2: *Which finite rings* $\mathbf{R}$ *with unit have the property that the variety of left unitary modules over* $\mathbf{R}$ *is decidable (*ω*-structured)?*

W. Baur [1976] constructed some finite rings having an undecidable theory of modules. F. Point [1986] and M. Prest [1988] have made a deep study of Problem 2 focused especially on hereditary rings.

We note that every variety of algebras of finite type is interpretable into the class of graphs. Thus a variety of finite type is unstructured just in case it is bi-interpretable with the class of graphs. We remark that a class of modules cannot be unstructured, because such a class is stable (see Baur [1975]). Any affine variety is bi-interpretable with a class of modules, hence cannot be unstructured.

Let us say that a variety $\mathcal{V}$ in a language $\mathsf{L} = (\Phi, \rho)$ is **finitely presentable** iff there exists a finite set of operation symbols $\Phi' \subseteq \Phi$, and a finite set of equations $\Gamma' \subseteq \mathrm{Th}(\mathcal{V})$ involving only the operation symbols in Φ', so that every equation in $\mathrm{Th}(\mathcal{V})$ that involves only symbols in Φ' is a consequence of Γ', and for every $f \in \Phi$ there exists an L-term τ involving only the operation symbols of Φ' such that $\mathcal{V} \models f(\bar{x}) \approx \tau(\bar{x})$.

COROLLARY 14.1 *Let* $\mathbf{A}$ *be a finite algebra. If* $\boldsymbol{V}(\mathbf{A})$ *is structured, then* $\boldsymbol{V}(\mathbf{A})$ *is finitely presentable.*

PROOF. Assume that $\boldsymbol{V}(\mathbf{A})$ is structured. Let $\boldsymbol{V}(\mathbf{A}) = \mathcal{V}_1 \otimes \mathcal{V}_2 \otimes \mathcal{V}_3$ be the decomposition given by Theorem 13.10. Thus $\mathbf{A} \cong \mathbf{A}_1 \times \mathbf{A}_2 \times \mathbf{A}_3$ where $\mathbf{A}_i \in \mathcal{V}_i$; and we have that $\mathcal{V}_i = \boldsymbol{V}(\mathbf{A}_i)$. It follows easily from Theorem 0.17 (iii) that the strongly Abelian variety $\mathcal{V}_1$ is finitely presentable. The finitely generated affine variety $\mathcal{V}_2$ is finitely presentable. This can be proved without difficulty by examining our analysis of the term operations of an affine algebra, given in the proof of Lemma 13.12. The finitely generated discriminator variety $\mathcal{V}_3$ is finitely presentable. This follows from two known results of general algebra: (1) A clone of operations on a finite set containing a ternary majority operation is finitely generated (thus Clo $\mathbf{A}_3$ is finitely generated); (2) any finitely generated discriminator variety of finite type is finitely axiomatizable. Now $\boldsymbol{V}(\mathbf{A})$ is the product of three finitely presentable varieties, implying that $\boldsymbol{V}(\mathbf{A})$ is finitely presentable. □

More generally, for a locally finite, but not necessarily finitely generated, variety $\mathcal{V}$ with a decomposition $\mathcal{V} = \mathcal{V}_1 \otimes \mathcal{V}_2 \otimes \mathcal{V}_3$ as in Theorem 13.10, the varieties $\mathcal{V}_1$ and $\mathcal{V}_2$ are necessarily finitely presentable and finitely generated, and $\mathcal{V}$ itself is finitely presentable (or finitely generated) if and only if $\mathcal{V}_3$ has the same property.

COROLLARY 14.2 *Let* $\mathbf{A}$ *be a finite algebra. The properties of undecidability and hereditary undecidability coincide for* $\mathcal{V} = \boldsymbol{V}(\mathbf{A})$. *Moreover, if* $\mathcal{V}$ *is undecidable but structured, then* $\mathcal{V}_2$ *is undecidable.*

PROOF. If $\mathcal{V}$ is unstructured, the conclusion follows immediately. So suppose that $\mathcal{V}$ is structured, and let

$$\mathcal{V} = \mathcal{V}_1 \otimes \mathcal{V}_2 \otimes \mathcal{V}_3$$

as Theorem 13.10 asserts. Now $\mathcal{V}$ is finitely presentable, by Corollary 14.1; and it follows from this that if $\mathcal{V}$ is undecidable then it is hereditarily undecidable. The strongly Abelian variety $\mathcal{V}_1$ is decidable (by Corollary 12.17— since $\mathcal{V}_1 \subseteq \mathcal{V}$ and $\mathcal{V}$ is structured). The finitely generated and finitely presentable discriminator variety $\mathcal{V}_3$ is decidable, by the theorem of Burris and Werner. Now, given the above decomposition, it follows that $\mathcal{V}$ is decidable iff all of the three subvarieties are decidable iff the affine variety $\mathcal{V}_2$ is decidable. □

COROLLARY 14.3 *There exists an algorithm which produces, given a finite algebra of finite type, a finite ring with unit such that the algebra generates a decidable variety iff the variety of left unitary modules over the ring is decidable.*

PROOF. We can use the algorithm described in Burris, McKenzie (Theorem 11.3), adding to it the test for decidability of a strongly Abelian variety contained in our Theorem 12.19. We describe the algorithm informally. Given a finite algebra $\mathbf{A}$ of finite type, we can effectively list all the ternary term operations of $\mathbf{A}$—i.e., the members of $\mathrm{Clo}_3\mathbf{A}$. Then we can effectively prune the list until it contains just the ternary term operations $t(x_1, x_2, x_3)$ that are decomposition operations of $\mathbf{A}$—i.e., for which there exists a decomposition

$$\mathbf{A} \cong \mathbf{A}_1^{(t)} \times \mathbf{A}_2^{(t)} \times \mathbf{A}_3^{(t)}$$

such that $\mathbf{A}_i \models t \approx x_i$ for $1 \leq i \leq 3$. This list is nonvoid, for there are trivial decomposition operations. Next, we search the list looking for an operation t for which $\mathbf{A}_1^{(t)}$ has the structure described in Lemma 11.8 and Theorem 11.9 (guaranteeing that $\boldsymbol{V}(\mathbf{A}_1^{(t)})$ is strongly Abelian), and for which $\mathbf{A}_2^{(t)}$ is affine, and $\mathbf{A}_3^{(t)}$ generates a discriminator variety. (The algebras $\mathbf{A}_i^{(t)}$ can be computed from $\mathbf{A}$ and t, and the required properties of these algebras can be effectively checked.)

If there does not exist such an operation t, then we terminate the algorithm and output a fixed finite ring $\mathbf{R}$ whose class of modules is undecidable.

By Theorem 13.10, $\boldsymbol{V}(\mathbf{A})$ is undecidable in this case. Assume, however, that we do find such a term t. Then we have

$$\boldsymbol{V}(\mathbf{A}) = \boldsymbol{V}(\mathbf{A}_1^{(t)}) \otimes \boldsymbol{V}(\mathbf{A}_2^{(t)}) \otimes \boldsymbol{V}(\mathbf{A}_3^{(t)})$$

and $\boldsymbol{V}(\mathbf{A})$ is decidable if and only if each of $\boldsymbol{V}(\mathbf{A}_i^{(t)})$ is decidable. Now $\boldsymbol{V}(\mathbf{A}_3^{(t)})$ is, in fact, decidable as we noted earlier. And we can effectively determine if $\boldsymbol{V}(\mathbf{A}_1^{(t)})$ is decidable, using the criterion expressed in Theorem 12.19. If $\boldsymbol{V}(\mathbf{A}_1^{(t)})$ proves not to be decidable, then we output the same fixed finite ring $\mathbf{R}$ as before and terminate the algorithm. In the remaining case, we output the ring of the variety $\boldsymbol{V}(\mathbf{A}_2^{(t)})$, which is described in the proof of Lemma 13.12 and is easily computable from the free algebra on two generators in this variety. According to Burris and McKenzie (Theorem 10.6) $\boldsymbol{V}(\mathbf{A}_2^{(t)})$ is decidable iff the class of left unitary modules over this ring is decidable. □

The methods we have used in this book are clearly applicable to the next problem, although new methods may also be needed.

PROBLEM 3: *Which locally finite quasivarieties (universal Horn classes) are undecidable (unstructured, ω-unstructured)?*

A. P. Zamyatin [1973], [1978b] has analyzed this problem for semigroups and rings, while R. McKenzie [1982b] has considered this problem for groups.

PROBLEM 4: *Which locally finite varieties are ω-unstructured? For which locally finite varieties is the class of finite members undecidable?*

In A. P. Zamyatin [1976], a list is given of all the varieties of rings whose class of finite members is decidable. Recently P. M. Idziak [1988], [1989a], [1989b] has characterized those finitely generated congruence distributive varieties of finite type whose class of finite members is decidable. He proves that such a variety must be congruence permutable and the congruence lattice of every subdirectly irreducible algebra in the variety must be linearly ordered. If either of these conditions fail then the variety is shown to be ω-unstructured.

We should like to point out some results on Problem 4 that are implicitly contained in this book. Notice that if $\mathcal{V}$ is ω-unstructured then the class of finite members of $\mathcal{V}$ is undecidable, in fact hereditarily undecidable. There exist unstructured varieties whose class of finite members is decidable. For example, the variety of monadic algebras (cylindric algebras of dimension 1) is such a class, and it is a locally finite discriminator variety. However, we know of no example of an ω-structured variety whose class of finite members is undecidable, or a structured variety which is ω-unstructured.

Among the fifteen different constructions we presented in proving Theorem 13.10, those of Lemmas 6.4, 13.11, 13.12 and all those in Part II yield not only that the variety in question is unstructured, but also that it is ω-unstructured. That is because in each case the construction produces, corresponding to any finite graph, a finite algebra. Thus we have the following corollary, which parallels Theorem 9.6.

COROLLARY 14.4 *Every ω-structured, locally finite, Abelian variety is the join of a strongly Abelian variety and an affine variety.*

PROOF. The proof of this result is obtained by modifying a few words in Chapters 7 through 10. □

Valeriote [1989] has strengthened this result to show that in fact every ω-structured, locally finite, Abelian variety is the varietal product of a strongly Abelian variety and an affine variety.

We conclude our work with the remark that using the results presented here, B. Hart and M. Valeriote have been able to count the number of non-isomorphic models in locally finite varieties; i.e., they have determined the possible infinite fine spectrum functions of these varieties, and correlated the spectrum function with algebraic properties of the variety. This work will be published in Hart and Valeriote [1989].

Bibliography

Baur, W. [1975]. $\aleph_0$-categorical modules, *Journal of Symbolic Logic* **40**, 213-220.

Baur, W. [1976]. Undecidability of the theory of Abelian groups with a subgroup, *Proc. Amer. Math. Soc.* **55**, 125-128.

Berman, J. and McKenzie, R. [1984]. Clones satisfying the term condition, *Discrete Math.* **52**, 7-29.

Burris, S. [1982]. The first order theory of Boolean algebras with a distinguished group of automorphisms, *Algebra Universalis* **15**, 156-161.

Burris, S. [1985], Iterated discriminator varieties have undecidable theories, *Algebra Universalis* **21**, 54-61.

Burris, S. and McKenzie, R. [1981]. *Decidability and Boolean Representations*, Memoirs Amer. Math. Soc. No. 246.

Burris, S., McKenzie, R. and Valeriote, M. [1989]. Decidable discriminator varieties (manuscript).

Burris, S. and Sankappanavar, H. P. [1981]. *A Course in Universal Algebra.* Graduate Texts in Mathematics, Springer-Verlag, New York.

Burris, S. and Werner, H. [1979]. Sheaf constructions and their elementary properties, *Trans. Amer. Math. Soc.* **248**, 269-309.

Ehrenfeucht, A. [1959]. Decidability of the theory of one function, *Notices Amer. Math. Soc.* **6**, p. 268.

Eklof, P. and Fisher, E. [1972]. The elementary theory of Abelian groups, *Ann. of Math. Logic* **4**, 115-171.

Ershov, Yu. L. [1972]. Elementary theories of groups, *Dokl. Akad. Nauk SSSR*, No. 6, 1240-1243.

Ershov, Yu. L., Lavrov, I. A., Taimanov, A. D. and Taitslin, M. A. [1965]. Elementary theories, *Russian Math. Surveys* **20**, 35-105.

Freese, R. and McKenzie, R. [1987]. *Commutator theory for congruence modular varieties*, London Math. Soc. Lecture Notes 125.

Gödel, K. [1931]. Über formal unentscheidbare Sätze der Principia Mathematica und verwandter Systeme I, *Monatshefte für Math. und Phys.* **38**, 173-198.

Grzegorczyk, A. [1951]. Undecidability of some topological theories, *Fund. Math.* **38**, 137-152.

Hart, B. and Valeriote, M. [1989]. The spectrum function for strongly abelian varieties (manuscript).

Hobby, D. and McKenzie, R. [1988]. *The Structure of Finite Algebras*, Amer. Math. Soc. Contemporary Mathematics Volume 76.

Idziak, P. M. [1988]. Reduced sub-powers and the decision problem for finite algebras in arithmetical varieties, *Algebra Universalis* **25**, 365-383.

Idziak, P. M. [1989a]. Varieties with decidable finite algebras I: linearity, *Algebra Universalis* **26**, 234-246.

Idziak, P. M. [1989b]. Varieties with decidable finite algebras II: permutability, *Algebra Universalis* **26**, 247-256.

Lavrov, I. A. [1963]. Effective inseparability of the sets of identically true formulae and finitely refutable formulae for certain elementary theories, *Algebra i Logika, Seminar* 2, vol. 1, 5-18.

Maltsev, A. I. [1954]. On the general theory of algebraic systems (Russian), *Mat. Sb.* (N.S.) **35**, 3-20.

McKenzie, R. [1982a]. Narrowness implies uniformity, *Algebra Universalis*, **15**, No. 1, 67-85.

McKenzie, R. [1982b]. Subdirect powers of non-Abelian groups, *Houston Journal of Mathematics* **8**, 389-399.

McKenzie, R. [1983]. Finite forbidden lattices, in *Universal Algebra and Lattice Theory*, Springer Lecture Notes No. 1004.

McKenzie, R., McNulty, G. and Taylor, W. [1987]. *Algebras, Lattices, Varieties*, Wadsworth and Brooks/Cole, Monterey, California.

Monk, J. D. [1976]. *Mathematical Logic*. Graduate Texts in Mathematics, Springer-Verlag, New York.

Point, F. [1986]. Problèmes de décidabilité pour les théories des modules, *Bull. Belg. Math. Soc. Ser. B* **38**, 58-74.

Prest, M. [1988]. *Model Theory and Modules*, London Math. Soc. Lecture Notes 130.

Rabin, M. O. [1965]. A simple method for undecidability proofs and some applications, in *Logic, Methodology and Philosophy of Science*, Bar-Hillel ed., North-Holland, Amsterdam.

Rabin, M. O. [1969]. Decidability of second-order theories and automata on infinite trees, *Trans. Amer. Soc.* **141**, 1-35.

Szmielew, W. [1955]. Elementary properties of Abelian groups, *Fund. Math.* **41**, 203-271.

Tarski, A. [1949]. Arithmetical classes and types of Boolean algebras, *Bull. Amer. Math. Soc.* **55**, p. 64.

Tarski, A. [1951]. *A decision method for elementary algebra and geometry*, The Rand Corporation, Santa Monica.

Tarski, A., Mostowski, A. and Robinson, R. M. [1953]. *Undecidable Theories*, North-Holland, Amsterdam.

Taylor, W. [1973]. Characterizing Mal'cev conditions, *Algebra Universalis* **3**, 351-397.

Trakhtenbrot, B. A. [1953]. On recursive separability, *DAN* **88**, 953-956.

Valeriote, M. [1986]. On decidable locally finite varieties, *PhD. dissertation, University of California, Berkeley.*

Valeriote, M. [1988]. Decidable unary varieties, *Algebra Universalis* **24**, 1-20.

Valeriote, M. [1989]. Abelian varieties having their class of finite algebras decidable (manuscript).

Zamyatin, A. P. [1973]. A prevariety of semigroups whose elementary theory is solvable, *Algebra and Logic* **12**, 233-241.

Zamyatin, A. P. [1976]. Varieties of associative rings whose elementary theory is decidable, *Soviet Math. Dokl.* **17**, 996-999.

Zamyatin, A. P. [1978a]. A non-Abelian variety of groups has an undecidable elementary theory, *Algebra and Logic* **17**, 13-17.

Zamyatin, A. P. [1978b]. Prevarieties of associative rings whose elementary theory is decidable, *Sib. Math. Zh.* **19**, 890-901.

Notation

Index

Progress in Mathematics

Progress in Mathematics is a series of books intended for professional mathematicians and scientists, encompassing all areas of pure mathematics. This distinguished series, which began in 1979, includes authored monographs and edited collections of papers on important research developments as well as expositions of particular subject areas.

All books in the series are "camera-ready", that is they are photographically reproduced and printed directly from a final-edited manuscript that has been prepared by the author. Manuscripts should be no less than 100 and preferably no more than 500 pages.

Proposals should be sent directly to the editors or to: Birkhäuser Boston, 675 Massachusetts Avenue, Suite 601, Cambridge, MA 02139, U.S.A.

1 GROSS. Quadratic Forms in Infinite-Dimensional Vector Spaces
2 PHAM. Singularités des Systèmes Différentiels de Gauss-Manin
3 OKONEK/SCHNEIDER/SPINDLER. Vector Bundles on Complex Projective Spaces
4 AUPETIT. Complex Approximation. Proceedings, Quebec, Canada, July 3–8, 1978
5 HELGASON. The Radon Transform
6 LION/VERGNE. The Weil Representation. Maslov Index and Theta Series
7 HIRSCHOWITZ. Vector Bundles and Differential Equations Proceedings, Nice, France, June 12–17, 1979
8 GUCKENHEIMER/MOSER/NEWHOUSE. Dynamical Systems, C.I.M.E. Lectures, Bressanone, Italy, June 1978
9 SPRINGER. Linear Algebraic Groups
10 KATOK. Ergodic Theory and Dynamical Systems I
11 BALSLEV. 18th Scandinavian Congress of Mathematicians, Aarhus, Denmark, 1980
12 BERTIN. Séminaire de Théorie des Nombres, Paris 1979–80
13 HELGASON. Topics in Harmonic Analysis on Homogeneous Spaces
14 HANO/MARIMOTO/MURAKAMI/OKAMOTO/OZEKI. Manifolds and Lie Groups: Papers in Honor of Yozo Matsushima
15 VOGAN. Representations of Real Reductive Lie Groups
16 GRIFFITHS/MORGAN. Rational Homotopy Theory and Differential Forms
17 VOVSI. Triangular Products of Group Representations and Their Applications
18 FRESNEL/VAN DER PUT. Géometrie Analytique Rigide et Applications
19 ODA. Periods of Hilbert Modular Surfaces
20 STEVENS. Arithmetic on Modular Curves
21 KATOK. Ergodic Theory and Dynamical Systems II
22 BERTIN. Séminaire de Théorie des Nombres. Paris 1980–81
23 WEIL. Adeles and Algebraic Groups

24 LE BARZ/HERVIER. Enumerative Geometry and Classical Algebraic Geometry
25 GRIFFITHS. Exterior Differential Systems and the Calculus of Variations
26 KOBLITZ. Number Theory Related to Fermat's Last Theorem
27 BROCKETT/MILLMAN/SUSSMAN. Differential Geometric Control Theory
28 MUMFORD. Tata Lectures on Theta I
29 FRIEDMAN/MORRISON. Birational Geometry of Degenerations
30 YANO/KON. CR Submanifolds of Kaehlerian and Sasakian Manifolds
31 BERTRAND/WALDSCHMIDT. Approximations Diophantiennes et Nombres Transcendants
32 BOOKS/GRAY/REINHART. Differential Geometry
33 ZUILY. Uniqueness and Non-Uniqueness in the Cauchy Problem
34 KASHIWARA. Systems of Microdifferential Equations
35 ARTIN/TATE. Arithmetic and Geometry: Papers Dedicated to I.R. Shafarevich on the Occasion of His Sixtieth Birthday, Vol. 1
36 ARTIN/TATE. Arithmetic and Geometry: Papers Dedicated to I.R. Shafarevich on the Occasion of His Sixtieth Birthday. Vol. II
37 DE MONVEL. Mathématique et Physique
38 BERTIN. Séminaire de Théorie des Nombres, Paris 1981–82
39 UENO. Classification of Algebraic and Analytic Manifolds
40 TROMBI. Representation Theory of Reductive Groups
41 STANELY. Combinatories and Commutative Algebra
42 JOUANOLOU. Théorèmes de Bertini et Applications
43 MUMFORD. Tata Lectures on Theta II
44 KAC. Infinite Dimensional Lie Algebras
45 BISMUT. Large Deviations and the Malliavin Calculus
46 SATAKE/MORITA. Automorphic Forms of Several Variables Taniguchi Symposium, Katata, 1983
47 TATE. Les Conjectures de Stark sur les Fonctions L d'Artin en s = 0
48 FRÖHLICH. Classgroups and Hermitian Modules
49 SCHLICHTKRULL. Hyperfunctions and Harmonic Analysis on Symmetric Spaces
50 BOREL, ET AL. Intersection Cohomology
51 BERTIN/GOLDSTEIN. Séminaire de Théoire des Nombres. Paris 1982–83
52 GASQUI/GOLDSCHMIDT. Déformations Infinitesimales des Structures Conformes Plates
53 LAURENT. Théorie de la Deuxième Microlocalisation dans le Domaine Complèxe
54 VERDIER/LE POTIER. Module des Fibres Stables sur les Courbes Algébriques Notes de l'Ecole Normale Supérieure, Printemps, 1983
55 EICHLER/ZAGIER. The Theory of Jacobi Forms
56 SHIFFMAN/SOMMESE. Vanishing Theorems on Complex Manifolds
57 RIESEL. Prime Numbers and Computer Methods for Factorization
58 HELFFER/NOURRIGAT. Hypoellipticité Maximale pour des Opérateurs Polynomes de Champs de Vecteurs
59 GOLDSTEIN. Séminarie de Théorie des Nombres, Paris 1983–84
60 PROCESI. Geometry Today: Giornate Di Geometria, Roma. 1984
61 BALLMANN/GROMOV/SCHROEDER. Manifolds of Nonpositive Curvature
62 GUILLOU/MARIN. A la Recherche de la Topologie Perdue
63 GOLDSTEIN. Séminaire de Théorie des Nombres, Paris 1984–85
64 MYUNG. Malcev-Admissible Algebras
65 GRUBB. Functional Calculus of Pseudo-Differential Boundary Problems

66 CASSOU-NOGUÈS/TAYLOR. Elliptic Functions and Rings and Integers
67 HOWE. Discrete Groups in Geometry and Analysis: Papers in Honor of G.D., Mostow on His Sixtieth Birthday
68 ROBERT. Antour de L'Approximation Semi-Classique
69 FARAUT/HARZALLAH. Deux Cours d'Analyse
70 ADOLPHSON/CONREY/GHOSH/YAGER. Number Theory and Diophantine Problems: Proceedings of a Conference at Oklahoma State University
71 GOLDSTEIN. Séminaire de Théories des Nombres, Paris 1985–1986
72 VAISMAN. Symplectic Geometry and Secondary Characteristics Classes
73 MOLINO. Reimannian Foliations
74 HENKIN/LEITERER. Andreotti–Grauert Theory by Integral Formulas
75 GOLDSTEIN, Séminaire de Théories des Nombres, Paris 1986–87
76 COSSEC/DOLGACHEV. Enriques Surfaces I
77 REYSAAT. Quelques Aspects des Surfaces de Riemann
78 BORHO/BRYLINSKI/MACPHERSON. Nilpotent Orbits, Primitive Ideals, and Characteristic Classes
79 MCKENZIE/VALERIOTE. The Structure of Decidable Locally Finite Varieties
80 KRAFT/PETRIE/SCHWARZ. Topological Methods in Algebraic Transformation Groups